Pseudo-Differential Operators
Theory and Applications
Vol. 10

Managing Editor

M.W. Wong (York University, Canada)

Editorial Board

Luigi Rodino (Università di Torino, Italy)
Bert-Wolfgang Schulze (Universität Potsdam, Germany)
Johannes Sjöstrand (Université de Bourgogne, Dijon, France)
Sundaram Thangavelu (Indian Institute of Science at Bangalore, India)
Maciej Zworski (University of California at Berkeley, USA)

Pseudo-Differential Operators: Theory and Applications is a series of moderately priced graduate-level textbooks and monographs appealing to students and experts alike. Pseudo-differential operators are understood in a very broad sense and include such topics as harmonic analysis, PDE, geometry, mathematical physics, microlocal analysis, time-frequency analysis, imaging and computations. Modern trends and novel applications in mathematics, natural sciences, medicine, scientific computing, and engineering are highlighted.

Vladimir Nazaikinskii • Bert-Wolfgang Schulze
Boris Sternin

The Localization Problem in Index Theory of Elliptic Operators

Vladimir Nazaikinskii
Ishlinsky Inst. f. Problems of Mechanics
Russian Academy of Sciences
Moscow
Russia

Bert-Wolfgang Schulze
Institut für Mathematik
Universität Potsdam
Potsdam
Germany

Boris Sternin
Peoples' Friendship Univers. of Russia
Moscow
Russia

With drawings by Olga Lazarenko

ISBN 978-3-0348-0509-4   ISBN 978-3-0348-0510-0 (eBook)
DOI 10.1007/978-3-0348-0510-0
Springer Basel Heidelberg New York Dordrecht London

Library of Congress Control Number: 2013953002

Mathematics Subject Classification (2010): 58J20, 19K56, 46L80, 35S35

© Springer Basel 2014

This work is subject to copyright. All rights are reserved by the Publisher, whether the whole or part of the material is concerned, specifically the rights of translation, reprinting, reuse of illustrations, recitation, broadcasting, reproduction on microfilms or in any other physical way, and transmission or information storage and retrieval, electronic adaptation, computer software, or by similar or dissimilar methodology now known or hereafter developed. Exempted from this legal reservation are brief excerpts in connection with reviews or scholarly analysis or material supplied specifically for the purpose of being entered and executed on a computer system, for exclusive use by the purchaser of the work. Duplication of this publication or parts thereof is permitted only under the provisions of the Copyright Law of the Publisher's location, in its current version, and permission for use must always be obtained from Springer. Permissions for use may be obtained through RightsLink at the Copyright Clearance Center. Violations are liable to prosecution under the respective Copyright Law.

The use of general descriptive names, registered names, trademarks, service marks, etc. in this publication does not imply, even in the absence of a specific statement, that such names are exempt from the relevant protective laws and regulations and therefore free for general use.

While the advice and information in this book are believed to be true and accurate at the date of publication, neither the authors nor the editors nor the publisher can accept any legal responsibility for any errors or omissions that may be made. The publisher makes no warranty, express or implied, with respect to the material contained herein.

Printed on acid-free paper

Springer Basel is part of Springer Science+Business Media (www.birkhauser-science.com)

Preface

This book deals with a localization approach to the index problem for elliptic operators. Localization ideas for many years have been widely used for solving various specific index problems, but the fact that there is actually a fundamental localization principle underlying all these solutions has mostly passed unnoticed. The ignorance of this general principle has often necessitated using various artificial tricks and hindered the solution of new important problems in index theory. So far, the general localization principle has been only scarcely covered in journal papers and not covered at all in monographs. The present book is intended to fill the gap. We explain the general localization principle and illustrate it by examples. The book is intended for working mathematicians as well as graduate and postgraduate university students specializing in differential equations and related topics.

In the construction of index formulas for elliptic operators on manifolds with boundary, singular manifolds, or noncompact manifolds with a special structure at infinity ("cylindrical ends"), the problem of separating index contributions from the "interior" part of the manifold and from the boundary, singular points, or a neighborhood of infinity is often important. Putting forward this problem is justified by the "locality" of the index. The fact that the index of an elliptic operator on a smooth compact manifold without boundary possesses some locality property was known in elliptic theory at least since the so-called "local index formulas" had emerged. A more careful consideration shows that locality property is actually a property not of the index itself, but of the *relative index*, i.e., the difference of indices of two operators differing on some subset of the manifold and coinciding elsewhere. For the case in which local index formulas are not known *a priori*, the proof of the locality property for the relative index is more complicated and has mostly been carried out on a case-by-case basis. For example, the locality property for the case of Dirac operators on complete noncompact Riemannian manifolds was proved by Gromov and Lawson, whose result was later generalized in various directions.

We consider a general functional-analytic model in which the locality principle holds for the relative index. One can refer to this principle more precisely as the *superposition principle for the relative index*. Note that the derivation of the superposition principle in our model is not based on any index formula, and hence the model applies in situations where index formulas are yet to be obtained. This abstract model serves as a source of relative index formulas (and, under additional assumptions like symmetry conditions, of index formulas) in various specific cases. By way of example, we present applications to the index of elliptic operators on noncompact manifolds and the index of elliptic boundary value problems and also briefly mention the index problem for elliptic operators (pseudodifferential operators and Fourier integral operators) on manifolds with singularities (which is treated in detail in another book by the authors). Furthermore, we include some recent results where the localization principle is used to compute the spectral flow of a family of Dirac operators on a manifold with boundary and this computation is applied to the description of the Aharonov–Bohm effect for massless Dirac fermions in graphene.

The outline of the book is as follows. The introduction exposes the superposition principle for the relative index at the most elementary level and briefly presents its applications, some of which are covered in detail in the main body of the book. It concludes with very brief bibliographical remarks, which are by no means exhaustive but do give some insight into the history of the subject and also indicate possible further reading.

Part I deals with the theory of the superposition principle. Chapter 1 introduces the general superposition principle for the relative index in the widest setting, and Chapters 2 and 3 provide a generalization of this principle to K-homology and Kasparov's KK-theory, thus putting the topic into the context of noncommutative geometry.

Part II contains examples of applications of the superposition principle to various specific problems, including those to elliptic operators on smooth manifolds (Chapter 4), boundary value problems (Chapter 5), and the spectral flow (Chapter 6).

Acknowledgements. The authors are keenly grateful to Victor Shatalov and Anton Savin for useful discussions. We also thank Olga Lazarenko, who has drawn all the figures.

Moscow–Potsdam–Hannover
2013

Vladimir Nazaikinskii
Bert-Wolfgang Schulze
Boris Sternin

Contents

Preface	v
Introduction	1
0.1 Basics of Elliptic Theory	1
0.2 Surgery and the Superposition Principle	3
0.3 Examples and Applications	9
0.4 Bibliographical Remarks	16

I Superposition Principle — 19

1 Superposition Principle for the Relative Index — 21
- 1.1 Collar Spaces — 21
- 1.2 Proper Operators and Fredholm Operators — 25
- 1.3 Superposition Principle — 29

2 Superposition Principle for K-Homology — 41
- 2.1 Preliminaries — 41
- 2.2 Fredholm Modules and K-Homology — 46
- 2.3 Superposition Principle — 48
- 2.4 Fredholm Modules and Elliptic Operators — 54

3 Superposition Principle for KK-Theory — 59
- 3.1 Preliminaries — 59
- 3.2 Hilbert Modules, Kasparov Modules, and KK — 59
- 3.3 Superposition Principle — 61

II Examples — 69

4 Elliptic Operators on Noncompact Manifolds — 71
- 4.1 Gromov–Lawson Theorem — 71
- 4.2 Bunke Theorem — 76

	4.3 Roe's Relative Index Construction	79
5	**Applications to Boundary Value Problems**	**81**
	5.1 Preliminaries	81
	5.2 Agranovich–Dynin Theorem	87
	5.3 Agranovich Theorem	89
	5.4 Bojarski Theorem and Its Generalizations	90
	5.5 Boundary Value Problems with Symmetric Conormal Symbol	91
6	**Spectral Flow for Families of Dirac Type Operators with Classical Boundary Conditions**	**93**
	6.1 Statement of the Problem	93
	6.2 Simple Example	97
	6.3 Formula for the Spectral Flow	100
	6.4 Computation of the Spectral Flow for a Graphene Sheet	108
Bibliography		**109**
Index		**115**

Introduction

The main subject of this book is the index locality principle, which can more precisely be called the *superposition principle for the relative index*. This introduction, which is partly based on the paper [61] by Nazaikinskii and Sternin, is intended as an elementary introduction to this principle. We discuss it and give some examples of its consequences and applications showing that it often proves to be a powerful tool for obtaining index formulas in various situations. Here we try to keep things as clear as possible and often give only the simplest versions of the results. A more detailed exposition, as well as the proofs, is given in subsequent chapters. Readers are also encouraged to consult the bibliographical remarks at the end of the introduction, which in particular recommend some further reading.

0.1 Basics of Elliptic Theory

We start by recalling elementary notions of elliptic theory. For an in-depth reading, we recommend [8, 64] and references cited therein.

Elliptic operators. Let M be a smooth compact manifold, and let D be a differential operator on M. In local coordinates, it can be written as

$$D = \sum_{|\alpha| \leq m} a_\alpha(x) \left(-i \frac{\partial}{\partial x}\right)^\alpha,$$

where m is the *order* of D. The *principal symbol* (characteristic polynomial) of D, given in local coordinates by the formula

$$\sigma(D) = \sum_{|\alpha| = m} a_\alpha(x) \xi^\alpha,$$

is a well-defined function on the cotangent bundle T^*M.

Remark 0.1. An operator D of order m can always be treated as an operator of order l for arbitrary $l \geq m$, and if $l > m$, then the principal symbol will be identically zero. So one in principle should speak more carefully and refer to $\sigma(D)$ defined above as the *principal symbol of D viewed as an operator of order m*. However, we omit these lengthier phrases because what is meant is always very clear from the context, and misunderstanding is unlikely to arise.

Definition 0.2. One says that an operator D is *elliptic* if the principal symbol $\sigma(D)$ is invertible everywhere on $T_0^*M = T^*M \setminus \{0\}$ (where $\{0\}$ stands for the zero section of the cotangent bundle). Plainly speaking, the principal symbol should be invertible for $\xi \neq 0$.

The following theorem describes one of the most important properties of elliptic operators.

Theorem 0.3 (Finiteness theorem). *If D is an elliptic differential operator of order m on a smooth compact manifold M, then it is Fredholm in the Sobolev spaces*

$$D : H^s(M) \longrightarrow H^{s-m}(M)$$

for every $s \in \mathbb{R}$. Moreover, the kernel of D (and the cokernel of D treated as the kernel of the L^2-adjoint operator D^*) are independent of s.

Recall that

- An operator D is said to be *Fredholm* if its kernel and cokernel are finite-dimensional; i.e., $\dim \ker D < \infty$ and $\dim \operatorname{coker} D < \infty$.

- The Sobolev spaces on M are defined in a standard manner (e.g., see [74]). In local coordinates, the H^s-norm can be written (up to equivalence) as

$$\|u\|_s = \left\{ \iint |\widetilde{u}|^2(\xi)(1+|\xi|^2)^s d\xi \right\}^{1/2},$$

where $\widetilde{u}(\xi)$ is the Fourier transform of $u(x)$.

Index. Let D be an elliptic operator of order m on M. The numbers $\dim \ker D$ and $\dim \operatorname{coker} D$ may change if we continuously perturb D in the class of elliptic operators, but their difference

$$\operatorname{ind} D = \dim \ker D - \dim \operatorname{coker} D$$

does not change under such perturbations. In other words, it is a *homotopy invariant*, and that is why it is of interest to study this difference. It is called the *index* of the operator D. In particular, the invariance under continuous perturbations in the class of elliptic operators implies that

- The index of D does not depend on lower-order terms in D and hence depends only on the principal symbol of D.

- Furthermore, the index of D depends only on the homotopy class of the principal symbol of D in the class of elliptic symbols (i.e., symbols invertible on T_0^*M).

0.2. Surgery and the Superposition Principle 3

The problem of computing the index of an elliptic operator D in terms of topological invariants of the principal symbol $\sigma(D)$, which was put forward by I. M. Gelfand, was solved by M. Atiyah and I. Singer in their famous index theorem [8].

However, the index problem is substantially more complicated for more general elliptic operators (in particular, for elliptic operators on singular or noncompact manifolds, for boundary value problems, for elliptic Fourier integral operators, etc.). These situations, which are not covered by the Atiyah–Singer theorem, require some new methods. In particular, any techniques that could assist one in solving the index problem, possibly reducing it (at least partly) to the case of smooth compact manifolds, are welcome. One such technique, which is the main subject of this book, is described in the next section.

0.2 Surgery and the Superposition Principle

There are quite a few methods that can be used to analyze and solve index problems. Here we deal with only one method, know as *surgery*. In this method, one uses cutting and pasting of manifolds (and accordingly of elliptic operators) to reduce complicated index problems to simpler ones. Cutting and pasting would be completely useless if we could not tell how the index of the operator in question changes under these operations. Fortunately enough, this change can be studied, and one can tell exactly what the change is. The answer is given by the *superposition principle*, which holds for the relative index (i.e., index increment) caused by surgery. In this section, we introduce the superposition principle in a simple example, then give a general statement of that principle, and finally present more examples, some of which will be discussed in full detail in subsequent chapters dealing with the applications of this principle.

0.2.1 Classical Example: Operators on Closed Manifolds

Let M be a compact C^∞ manifold without boundary, and let D be an elliptic differential operator on M. Further, let disjoint closed subsets A and B of the manifold M be given; an example is shown in Fig. 0.1. Now let us make some surgery on A. From M, we cut away the piece A and replace it by some other (smooth) piece A' (see Fig. 0.2). Next, we extend the operator D from $M \setminus A$ to the entire new manifold

$$M_A = (M \setminus A) \cup A'$$

so as to obtain a new elliptic operator, D_A. (For the sake of argument, let us assume that this can be done.) The index of D_A is in general different from the index of D, and we obtain the index increment (relative index)

$$\triangle_A \stackrel{\text{def}}{=} \operatorname{ind} D_A - \operatorname{ind} D.$$

Likewise, let us make a surgery on B by cutting away the piece B and replacing it by some piece B' (see Fig. 0.3) and then extending the operator D from $M \setminus B$ to the manifold

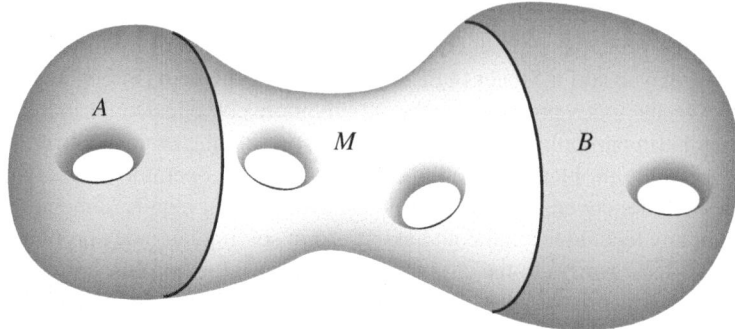

Figure 0.1: Manifold M

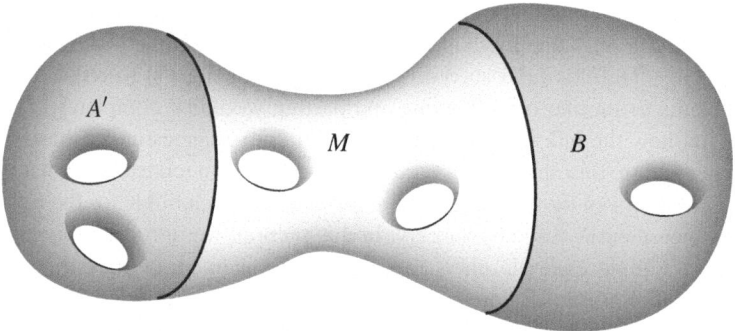

Figure 0.2: Surgery on A

$$M_B = (M \setminus B) \cup B'$$

so as to obtain a new elliptic operator, D_B. (Again, assume that this can be done.) Now we have the new index increment

$$\triangle_B \stackrel{\text{def}}{=} \operatorname{ind} D_B - \operatorname{ind} D.$$

These two surgeries are totally independent: when modifying the operator D over A, we do not touch anything away from A (in particular, on B), and vice versa. Now we can apply both modifications (surgeries) simultaneously, and the result for the operator will be the same as if we applied first one surgery and then the other, their order being irrelevant (see Fig. 0.4). The resulting operator will be denoted by $D_{A \cup B}$ and the index increment by

$$\triangle_{A \cup B} \stackrel{\text{def}}{=} \operatorname{ind} D_{A \cup B} - \operatorname{ind} D.$$

It is natural to ask how this "total" increment is related to the "partial" increments $\triangle_A$ and $\triangle_B$. The answer is exactly as it should be.

0.2. Surgery and the Superposition Principle

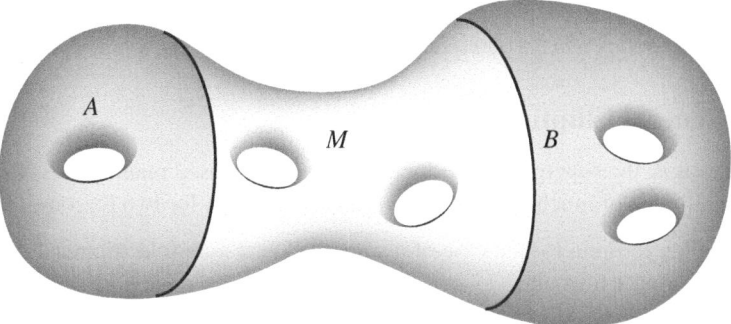

Figure 0.3: Surgery on B

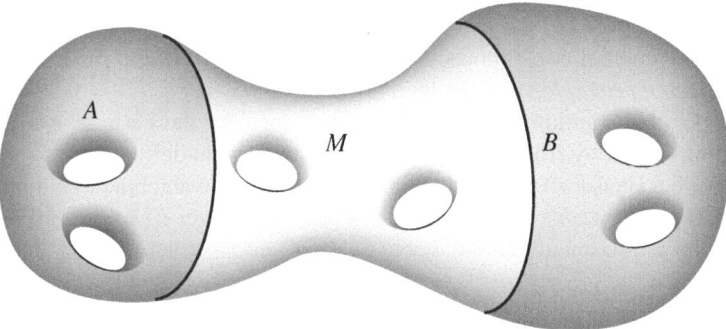

Figure 0.4: Surgery on A and B simultaneously

Lemma 0.4 (Superposition principle).
$$\triangle_{A\cup B} = \triangle_A + \triangle_B.$$

Proof. This follows from the *local index formula* (e.g., see [31])

$$\operatorname{ind} D = \int_M \alpha(x), \tag{1}$$

where the *local index density* $\alpha(x)$ at a point x depends only on $\sigma(D)$ and its derivatives in the fiber T_x^*M. Indeed, it suffices to note that, say,

$$\triangle_A = \int_A (\alpha(x) - \alpha'(x)),$$

where α' is the local index density corresponding to D_A, since $\alpha = \alpha'$ outside A. The desired formula follows, since the integral is an additive set function. □

Remark 0.5. The superposition principle means that index increments stemming from independent surgeries behave additively.

0.2.2 General Elliptic Operators

Beautiful as it is, the superposition principle on smooth closed manifolds is, for two reasons, generally not of much help when computing the index for two reasons.

1. The index formula for the case of a smooth closed manifold is already known (the Atiyah–Singer formula).

2. The proof given above is hardly satisfactory, since it relies on the fact that we already know the (local) index formula.

Hence we wish to generalize this principle to cases beyond the Atiyah–Singer theorem and, moreover, invent a proof that does not rely on the presence of any *a priori* known index formula.

Thus the problem is as follows: *Describe a sufficiently general framework in which the superposition principle for index increments is valid.*

One possible solution to this problem is to consider the class of *general elliptic operators* introduced by Atiyah [4]. Let us recall the relevant definitions.

Let X be a Hausdorff compactum, and let $C(X)$ be the algebra of continuous functions on X. Further, let H_1 and H_2 be Hilbert ∗-modules over $C(X)$, i.e., Hilbert spaces equipped with a ∗-action of the C^*-algebra $C(X)$.

Definition 0.6. An operator
$$A : H_1 \longrightarrow H_2$$
is called a general elliptic operator if the following two conditions are satisfied:

- A is Fredholm.

- A almost commutes with the action of $C(X)$:
$$\varphi A - A\varphi \in \mathscr{K}(H_1, H_2) \quad \forall \varphi \in C(X),$$
where $\mathscr{K}(H_1, H_2)$ is the set of compact linear operators from H_1 to H_2.

To state the superposition principle, we should first define the notion of surgery for general elliptic operators. This is however intuitively clear. Let D_1 and D_2 be two general elliptic operators (with the same underlying compactum X). Next, let $A \subset X$ be a closed set.

Definition 0.7. We say that D_1 and D_2 are obtained from each other by a *modification* (or *surgery*) on A if for each function $\varphi \in C(X)$ whose support does not meet A (i.e., $\operatorname{supp} \varphi \cap A = \varnothing$) one has
$$\varphi D_1 \varphi \equiv \varphi D_2 \varphi \quad \text{modulo compact operators.}$$

In this case, we write $D_1 \xrightarrow{A} D_2$ or $D_2 \xrightarrow{A} D_1$.

0.2. Surgery and the Superposition Principle

This definition, however, needs further clarification: we have not assumed that D_1 and D_2 act in the same spaces, so how can we compare $\varphi D_1 \varphi$ and $\varphi D_2 \varphi$? Let us give necessary explanations (which prove to be a bit technical).

If H is a Hilbert space equipped with an action of $C(X)$, then the notion of support $\operatorname{supp} u \subset X$ is well defined for all $u \in H$ in a natural way: a point x does *not* belong to the support of u if $\varphi u = 0$ for all $\varphi \in C(X)$ supported in a sufficiently small neighborhood of x.

We introduce the following notation: by $H_K \subset H$ we denote the closure of the set of elements $u \in H$ supported in K. Now if

$$D_1 : H_1 \longrightarrow G_1, \quad D_2 : H_2 \longrightarrow G_2$$

are general elliptic operators, then it makes sense to say that they are obtained from each other by a surgery on A if some isomorphisms

$$H_{1U} \longrightarrow H_{2U}, \quad G_{1U} \longrightarrow G_{2U}$$

of $*$-modules over $C(X)$, where $U = X \setminus A$ is the complement of A, are given and fixed.

Definition 0.8 (Definition of surgery on A revisited). We write

$$D_1 \xrightarrow{A} D_2$$

if the diagram

$$\begin{array}{ccc} H_{1U} & \xrightarrow{\varphi D_1 \varphi} & G_{1U} \\ \downarrow & & \downarrow \\ H_{2U} & \xrightarrow{\varphi D_2 \varphi} & H_{2U} \end{array}$$

commutes modulo compact operators for each $\varphi \in C(X)$ such that

$$\operatorname{supp} \varphi \cap A = \varnothing.$$

With this definition of surgery for general elliptic operators, the following superposition theorem holds for the index increments.

Theorem 0.9. *Let*

$$\begin{array}{ccc} D & \xrightarrow{A} & D_A \\ B \downarrow & & \downarrow B \\ D_B & \xrightarrow{A} & D_{AB} \end{array}$$

be a commutative diagram[1] of independent surgeries ($A \cap B = \varnothing$) of general elliptic operators over $C(X)$. Then

$$\triangle_{AB} = \triangle_A + \triangle_B.$$

The proof of this theorem will be given in Section 2.2.

---

[1] A diagram of surgeries is said to be commutative if the underlying isomorphisms of Hilbert spaces over $X \setminus (A \cup B)$ form a commutative diagram.

0.2.3 Operators in Collar Spaces

The theorem given in the preceding subsections does not cover applications related to Fourier integral operators (which do not almost commute with multiplication by functions). Furthermore, strictly speaking, it applies only to zero-order operators, since operators of positive order (in particular, any differential operators) do not compactly (or even boundedly) commute with continuous functions. So it is a good idea to devise a slightly different framework for the superposition principle, including the preceding as a special case.

This was done in [59], and here we describe the corresponding results very briefly. (See more details in Sections 1.1. and 1.2.) The main ideas of the approach are as follows.

1. We actually do not need arbitrary spaces X in applications of the superposition principle. If X is a compactum and $A, B \subset X$ are closed disjoint subsets, then there always exists a continuous mapping

$$f : X \longrightarrow [-1,1]$$

such that $A \subset f^{-1}(-1)$ and $B \subset f^{-1}(1)$. The mapping f induces the structure of a $C([-1,1])$-module on every $C(X)$-module, and so we can always assume that $X = [-1,1]$, $A = \{-1\}$, and $B = \{1\}$.

2. Instead of $C([-1,1])$-modules, one considers $C^\infty([-1,1])$-modules (for brevity referred to as *collar spaces*), which permits one to cover the case of positive-order operators (in particular, differential operators).

3. Finally, instead of single operators one considers families of operators depending on a small parameter such that the "support of the kernel" for these operators tends to the diagonal as the parameter tends to zero. (See the precise definitions in Section 1.1.) Thus, for each given parameter value, the operators need not be local; they are only "local in the limit." This permits one to consider a wider class of operators, including Fourier integral operators on manifolds with singularities, while the superposition principle remains true.

With these improved definitions, the main theorem remains valid.

Theorem 0.10 ([58, 59]). *Suppose that the following commutative diagram of modifications of elliptic operators in collar spaces holds:*

$$\begin{array}{ccc} D & \xleftrightarrow{-1} & D_- \\ {\scriptstyle 1}\updownarrow & & \updownarrow{\scriptstyle 1} \\ D_+ & \xleftrightarrow{-1} & D_\pm. \end{array}$$

Then

$$\mathrm{ind}(D) - \mathrm{ind}(D_-) = \mathrm{ind}(D_+) - \mathrm{ind}(D_\pm).$$

The proof will be given in Section 1.3.

0.2.4 Superposition Principle for K-Homology and KK-Theory

The preceding theorems deal with the relative index of elliptic operators. However, the index is by no means a unique homotopy invariant of elliptic operators: the group $\text{Ell}(M)$ of stable homotopy classes of elliptic operators on a manifold M is usually much richer than $\mathbb{Z}$. Thus it is natural to try to generalize the relative index superposition principle so as to cover the other invariant as well. This is done in Chapters 2 and 3, where we show how the elements of the K-homology group of a C^*-algebra (or, even more generally, of a Kasparov KK-group) behave under cutting and pasting. This is a very natural generalization, because, say, there is a natural construction that assigns an element of the K-homology of an algebra A to a general elliptic operator over A in the sense of Atiyah [4]. For a smooth closed manifold M, this construction gives the isomorphism $\text{Ell}(M) \simeq K^0(C(M))$. However, we do not restrict ourselves to commutative algebras $A = C(M)$ but state and prove the corresponding theorems for general algebras A in the spirit of noncommutative geometry [26].

0.3 Examples and Applications

Now let us give some examples of how the superposition principle works. These examples come from the following areas:

- Elliptic operators on manifolds with singularities.
- Elliptic operators on noncompact manifolds.
- Boundary value problems.
- Fourier integral operators.

0.3.1 Elliptic Operators on Manifolds with Conical Singularities

Let M be a manifold with a conical point a and base Λ of the cone (see Fig. 0.5). (For precise definitions, e.g., see [48].)

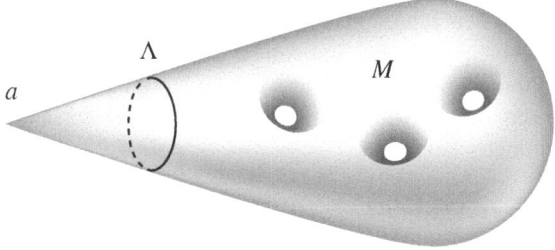

Figure 0.5: A manifold with conical singularities

Cone-degenerate differential operators on M near the conical point have the form of finite sums

$$D = \sum a_{\alpha j}(\omega, r) \left(-i\frac{\partial}{\partial \omega}\right)^\alpha \left(ir\frac{\partial}{\partial r}\right)^j,$$

where r is the distance from the conical point and ω is a coordinate on the base of the cone. The operator family

$$\sigma_c(D) = \sum a_{\alpha j}(\omega, 0) \left(-i\frac{\partial}{\partial \omega}\right)^\alpha p^j.$$

on Λ is called the *conormal symbol* of D. Cone-degenerate operators are considered in weighted Sobolev spaces $H^{s,\gamma}(M)$ (where the exponent s corresponds to smoothness and γ indicates that the weight r^γ is used near the conical point), and the elipticity condition for cone-degenerate operators is that the interior principal symbol is invertible outside the zero section of the ("compressed" [45]) cotangent bundle of M minus the zero section and the conormal symbol is invertible on the weight line $\operatorname{Im} p = \gamma$. (In what follows, we for simplicity always assume that $\gamma = 0$.)

Now let M be a manifold with conical point a, and let D be an elliptic operator on M. Consider the problem of finding the index $\operatorname{ind} D$. This problem can be solved with the help of the superposition principle for the relative index under certain symmetry conditions on the interior principal symbol of D. Namely, the following theorem holds.

Theorem 0.11 ([48]). *Suppose that, in some neighborhood of the conical point, the interior principal symbol of D satisfies the symmetry condition*

$$\sigma(D)(\omega, r, q, -p) = f_1 \sigma(D)(\omega, r, q, p) f_2,$$

where f_1 and f_2 are bundle isomorphisms on M in the above-mentioned neighborhood of the conical point. Then

$$\operatorname{ind} D = \frac{1}{2}(\operatorname{ind} 2D + \operatorname{ind} D_c),$$

where $2D$ is the elliptic operator on the double of M whose principal symbol is obtained by clutching with the use of symmetry conditions and D_c is an operator on the suspension $S\Lambda$ explicitly constructed from the conormal symbol of D.

Thus, the index of D is represented as the sum of two terms, one of which depends only on the interior principal symbol and can be expressed by the Atiyah–Singer index theorem (applied to the operator $2D$ on the compact closed manifold $2M$) and the other depends only on the conormal symbol. This theorem was first obtained in [77] under the slightly stronger symmetry condition

$$\sigma_c(D)(p) = f_1 \sigma_c(D)(p_0 - p) f_2$$

(where f_1 and f_2 are bundle automorphisms) imposed on the conormal rather than interior symbol. In this case, the second term in the index formula can be represented as a sum of multiplicities of poles of the operator family $\sigma_c(D)(p)^{-1}$ in a certain strip in the complex

0.3. Examples and Applications

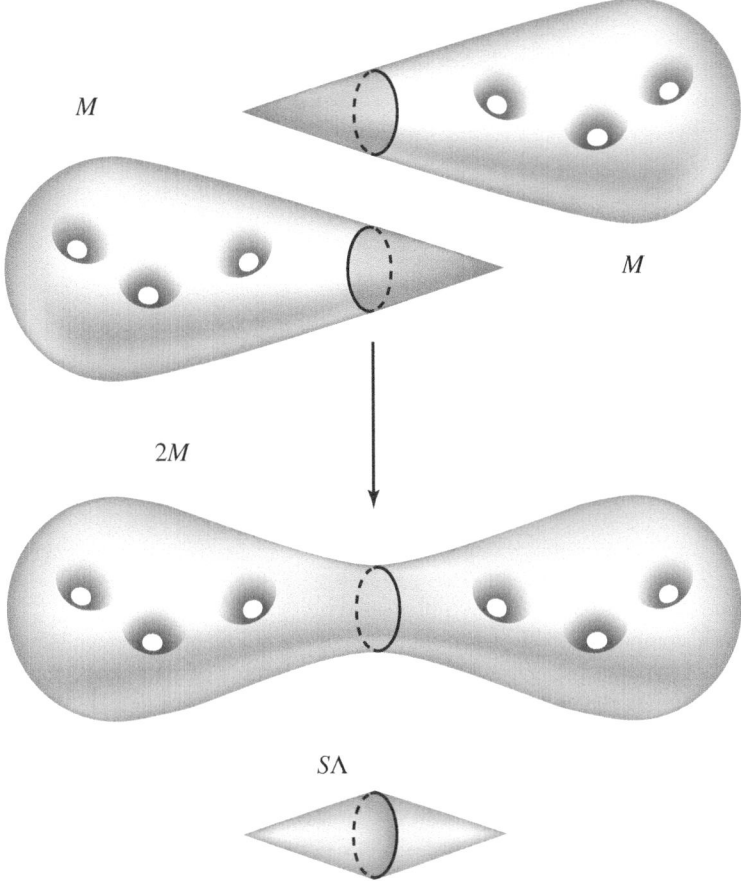

Figure 0.6: Surgery proving the index theorem for manifolds with conical singularities

plane. In the general case, the second term is expressed as the spectral flow of some homotopy of $\sigma_c(D)(p)$ to $f_1\sigma_c(D)(p_0 - p)f_2$. The proof of this index formula is given by surgery in conjunction with the superposition principle; the nontrivial part of the surgery diagram is shown in Fig. 0.6.

0.3.2 Elliptic Operators on Noncompact Manifolds

Let X_0 and X_1 be noncompact manifolds, and let D_0 and D_1 be Fredholm elliptic operators in certain L^2 spaces on these manifolds. Suppose that these manifolds coincide at infinity. Namely, there are compact sets $K_j \subset X_j$ and a measure-preserving diffeomorphism

$$X_0 \setminus K_0 \stackrel{a}{\simeq} X_1 \setminus K_1$$

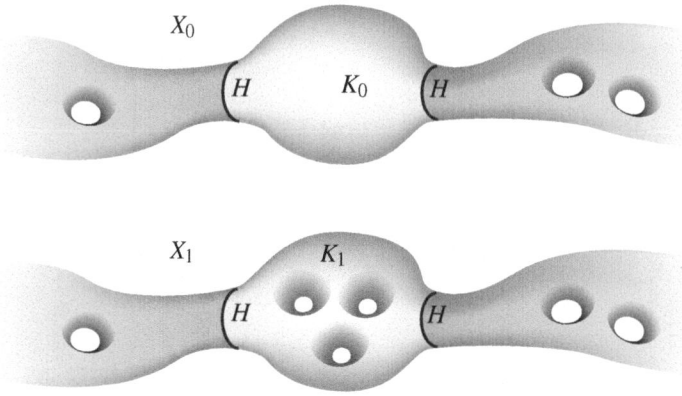

Figure 0.7: Noncompact manifolds coinciding at infinity

such that
$$D_1 = \Phi D_0 \Psi^{-1},$$
where Φ and Ψ vector bundle isomorphisms over a (see Fig. 0.7).

We can compactify both manifolds by cutting the noncompact ends away along some compact hypersurface H and then glueing the same compact "cap" to both manifolds and continuing the operators in the same way to the cap.

The new elliptic operators on the new compact manifolds $\widetilde{X}_0$ and $\widetilde{X}_1$ thus obtained will be denoted by $\widetilde{D}_0$ and $\widetilde{D}_1$.

The superposition theorem implies the following assertion.

Theorem 0.12. *One has*
$$\operatorname{ind} D_1 - \operatorname{ind} D_0 = \operatorname{ind} \widetilde{D}_1 - \operatorname{ind} \widetilde{D}_0.$$

This theorem was proved for the special case of Dirac operators on complete Riemannian manifolds by Gromov and Lawson [32] and later extended to a more general class of operators by Anghel [3]. See Section 4.1 for details.

0.3.3 Boundary Value Problems

The superposition principle for the relative index also has well-known manifestations in boundary value problems.

Let M be a compact manifold with boundary ∂M, and let D be an elliptic operator on M; in local coordinates near the boundary,
$$D = \sum_{|\alpha|+\beta \leq m} a_{\alpha\beta}(x,t) \left(-i\frac{\partial}{\partial x}\right)^{\alpha} \left(-i\frac{\partial}{\partial t}\right)^{\beta},$$

0.3. Examples and Applications

where we assume that the boundary is given by the equation $\partial M = \{t = 0\}$ and the interior of the manifold corresponds to positive t.

Consider a classical boundary value problem of the form

$$\begin{cases} Du = f, \\ Bu\big|_{\partial M} = g. \end{cases}$$

For short, we denote such a problem by (D, B).

Recall the ellipticity conditions for the problem (D, B). To obtain these conditions, one freezes the coefficients of the equation at some point $(x, 0) \in \partial M$, drops away lower-order terms, and makes the Fourier transform with respect to the variables tangent to the boundary. Thus we obtain the ordinary differential operator

$$\widetilde{D}(x, \xi) = \sum_{|\alpha| + \beta = m} a_{\alpha\beta}(x, 0) \xi^\alpha \left(-i \frac{\partial}{\partial t} \right)^\beta$$

with constant coefficients on the half-line $\mathbb{R}_+$. This operator depends on the parameters $(x, \xi) \in T_0^* \partial M$.

Let $L_+ \equiv L_+(x, \xi)$ be the subspace of initial data at $t = 0$ for solutions of $\widetilde{D}v = 0$ decaying as $t \to \infty$.

Condition 0.13 (Shapiro–Lopatinskii). $\sigma(B)|_{L_+}$ is an isomorphism for $\xi \neq 0$.

The main analytic theorem of the theory of boundary value problems is as follows.

Theorem 0.14. *If the boundary value problem (D, B) satisfies the Shapiro–Lopatinskii condition, then it is Fredholm.*

Now we are in a position to state two relative index theorems for boundary value problems.

Let D_1 and D_2 be two elliptic operators coinciding near ∂M, and let B be a boundary operator satisfying the Shapiro–Lopatinskii conditions with respect to one (and hence both) of the operators.

The superposition principle implies the following assertion.

Theorem 0.15. *One has*

$$\mathrm{ind}(D_1, B) - \mathrm{ind}(D_2, B) = \mathrm{ind}\, D$$

where D is a zero-order elliptic operator on M such that

$$\sigma(D) = \sigma(D_1)\sigma(D_2)^{-1}$$

and D acts as a vector bundle isomorphism near ∂M.

Another relative index theorem deals, on the opposite, with the case of one operator and two boundary conditions.

Let D be an elliptic operator on M, and let B_1 and B_2 be two boundary operators satisfying the Shapiro–Lopatinskii condition.

Then it follows from the superposition principle that the theorem below holds.

Theorem 0.16. *One has*

$$\mathrm{ind}(D, B_1) - \mathrm{ind}(D, B_2) = \mathrm{ind}\, C,$$

where C is an elliptic operator on ∂M with

$$\sigma(C) = \sigma(B_1)|_{L_+} (\sigma(B_2)|_{L_+})^{-1}.$$

These two theorems are known as the Agranovich and Agranovich–Dynin theorems (see [1, 2]). Surgery, in conjunction with the superposition principle, provides new, elementary proofs (see Chapter 5).

0.3.4 Fourier Integral Operators

The index problem for quantized canonical transformations (Fourier integral operators) was posed by Weinstein [81, 82], and its solution was obtained for smooth manifolds by Epstein and Melrose [29] in a particular case and by Leichtnam, Nest, and Tsygan [41] in the general case. The superposition principle permits one to derive an index formula for quantized contact transformations on singular manifolds. This was done in [53].

Let us briefly describe the construction of Fourier integral operators on a manifold with conical singularities. Let M be a manifold with a conical singular point α and base Λ of the cone. (We assume for simplicity that there is only one conical point.) We shall use the cylindrical model, i.e., pass from the coordinate r to the cylindrical coordinate t by the formula $r = e^{-t}$ (see Fig. 0.8).

Quantized canonical transformations are obtained by the quantization of classical transformations, so let us say a few words about the latter. We shall quantize homogeneous canonical (contact) transformations

$$g : T_0^* M \longrightarrow T_0^* M.$$

Transformations associated with the conical structure should be continuous up to $r = 0$. In the t-coordinate, this corresponds to "exponential stabilization of the coefficients" as $t \to \infty$. For simplicity, we impose an even stronger condition that the coefficients "are independent of t for sufficiently large t." Stated precisely, this means the following:

Condition 0.17 (Stabilization). The transformation g commutes with translations along the t-axis for $t \gg 0$.

In other words,

$$g|_{t \gg 0} = g_\infty : T_0^* C \longrightarrow T_0^* C,$$

0.3. Examples and Applications

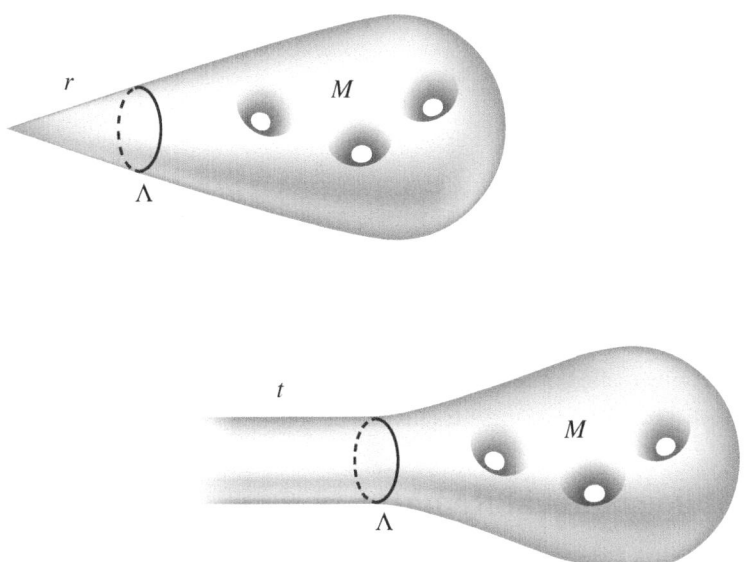

Figure 0.8: Cylindrical coordinates

where $C = \mathbb{R} \times \Lambda$ is the infinite cylinder with base Λ and g_∞ commutes with translations.

The quantized transformation is given by the Fourier integral operator associated with the graph $L_g \subset T_0^*M \times T_0^*M$ of the classical transformation g. This graph is a Lagrangian manifold, and we make the following assumption.

Assumption 0.18. The quantization condition (e.g., see [44, 46]) is satisfied for L_g (i.e., the *Maslov index* is zero on L_g).

Then the quantized canonical transformation

$$T_g : L^2(M) \longrightarrow L^2(M)$$

is defined in the usual manner as the Fourier integral operator with amplitude 1 associated with the Lagrangian manifold L_g. To ensure appropriate behavior near the conical point, we require that T_g commutes with translations along the t-axis for large t. This can be done in view of the similar condition imposed on g. We shall assume that T_g is elliptic.

Now we are in a position to state the index theorem. We impose the simplest symmetry condition on the classical transformation.

Condition 0.19. The transformation g_∞ commutes with the inversion $(t, p) \mapsto (-t, -p)$.

Then (in fact, after some homotopies) two copies of g can be glued into a canonical transformation $2g : T_0^*2M \longrightarrow T_0^*2M$ of the double of the cotangent bundle T_0^*M.

Surgery and the superposition principle give the following formula for the index of the quantized canonical transformation.

Theorem 0.20. *The index of an elliptic quantized canonical transformation is given by the formula*
$$\operatorname{ind} T_g = \frac{1}{2}(\operatorname{ind} T_{2g} + \operatorname{ind} T_{g_\infty}),$$
where

- T_{2g} *is a quantized canonical transformation on the smooth closed manifold* $2M$.
- T_{g_∞} *is a quantized canonical transformation on the cylinder* C.

Remark 0.21. 1. The index $\operatorname{ind} T_{2g}$ can be computed according to the index theorems given in [29, 41].

2. In special cases, $\operatorname{ind} T_{g_\infty}$ can be computed as the sum of multiplicities of poles of an operator family (called the conormal symbol) associated with T_g.

A detailed exposition of these results can be found in [48, 53].

0.4 Bibliographical Remarks

In closing, let us very briefly discuss the literature on the topic, indicate the sources of the results presented in the book, and suggest some further reading.

Relative index theorems, which deal with what happens with the index of elliptic operators when cutting and pasting, are abundant in the literature. The famous Gromov–Lawson relative index theorem [32] for Dirac operators on complete noncompact Riemannian manifolds is apparently the first relative index theorem in the literature. The results obtained by Gromov and Lawson were later generalized in various directions. For example, Anghel [3] generalized them to the case of arbitrary self-adjoint first-order elliptic operators on complete Riemannian manifolds. Donnelly [28] (for the signature operator) and Bunke [21] (for arbitrary Dirac operators) obtained "local" relative index theorems in terms of heat kernels. Lesch [42] obtained local relative index theorems for a class of symmetric first-order elliptic operators in the equivariant case; his setting also admits the presence of conical singularities in the interior of the manifold. We also mention the important book [16] by Booß-Bavnbek and Wojciechowski, where some relative index theorems for Dirac type operators can be found, and Teleman's paper [80], where relative index theorems were used in the context of the index problem for signature operators on Lipschitz manifolds. Bunke [22] considered Dirac operators on a complete noncompact Riemannian manifold in section spaces of bundles of projective Hilbert B-modules (e.g., see [47]), where B is a C^*-algebra, and obtained a relative index theorem for such operators, the index being an element of the K-group of B. In the same paper, Bunke applied these ideas to index computations for Callias type operators (i.e., operators on noncompact manifolds having the form $D+\Phi$ at infinity, where D is a Dirac type operator and Φ is a self-adjoint bundle endomorphism). These operators originally arose in Callias' papers [24, 25] and were later considered by Bott and Seeley [18] and Kottke [40]. The relative index on orbifolds was studied in [30].

Roe [70] put the relative index construction in the context of operator algebras introduced in [69] for studying index theory on noncompact manifolds and showed that

0.4. Bibliographical Remarks

this construction is equivalent to those considered by Gromov and Lawson as well as by Borisov, Müller, and Schrader [17] and Julg [37]. For further development of this line of research and related papers, see [34, 71–73] as well as [85–88].

Analogs of relative index theorems can also be proved for invariants of other kinds. Here we only mention the papers by Loya and Park [43], where spectral invariants were considered, higher relative index theorems in [83, 84] by Xie and Yu, and the study of the zeta determinant by Park and Wojciechowski [65].

Let us mention the recent papers [9] by Ballmann, Brüning, and Carron, where the index theory on manifolds with straight ends is studied, and [10] by Bär and Ballmann; in both papers, relative index theorems play an important role.

Let us indicate the sources of some of the results presented in the book.

The general relative index superposition principle (Chapter 1) is due to Nazaikinskii and Sternin [55–61]. The superposition principle for K-homology (Chapter 2) and KK-theory (Chapter 3) was obtained by Nazaikinskii in [63] and [62], respectively. The index theorems for elliptic operators on manifolds with conical singularities (Section 0.3.1) were originally obtained by Schulze, Sternin, and Shatalov [77] and later developed in a number of papers. The relevant references can be found in the book [48] by Nazaikinskii, Savin, Schulze, and Sternin, in which the application of the index superposition principle to index theory on manifolds with singularities (not necessarily isolated) is put in systematic form. The index theorem for Fourier integral operators on manifolds with conical singularities (Section 0.3.4) was obtained by Nazaikinskii, Schulze, and Sternin [53]. The results in Chapter 6 on the spectral flow of Dirac type operators with local boundary conditions on a compact manifold with boundary were obtained by Katsnelson and Nazaikinskii [38]. The results in other chapters and sections not explicitly mentioned here are due to numerous authors cited there.

Finally, note that neither these bibliographical remarks nor the bibliography itself is comprehensive. They merely reflect the authors' personal taste, and there are surely many important papers that are not on the list.

Part I
Superposition Principle

Chapter 1

Superposition Principle for the Relative Index

1.1 Collar Spaces

The superposition principle outlined in the introduction deals with modifications (surgeries) of elliptic operators performed on two disjoint closed subsets, say A and B, of the manifold M on which these operators are defined and says that the index increments $\triangle_A$ and $\triangle_B$ resulting from these surgeries are independent (i.e., are just summed if both surgeries are carried out simultaneously). In this chapter, we prove the superposition principle for the relative index in general form as it was stated in [55–60].

In fact, there are only two things important for the superposition principle to hold:

1. The two "regions" where the operators are to be modified should be *separated* from each other.

2. The operators in question should respect the separated subsets, i.e., be *local* in some sense.

The notion of collar spaces discussed in this section takes care of the first thing, and the second thing—the relevant class of operators—will be introduced in the next section.

Collar spaces were originally defined in [55–57,59]. These spaces are Hilbert spaces equipped with the structure of a module over the unital algebra $C^\infty([-1,1])$ of smooth functions on the closed interval $[-1,1]$. We will use the Fréchet topology defined on this algebra by the seminorms

$$\|f\|_k = \max_{t\in[-1,1]} |f^{(k)}(t)|.$$

Let us give a formal definition.

Definition 1.1. Let H be a separable Hilbert space, and let a continuous action of the algebra $C^\infty([-1,1])$ on H be defined. (We assume that the function $f(x) \equiv 1$ acts as the identity operator on H.) Then we say that H is a *collar space*.

Let us produce some examples of collar spaces.

Example 1.2. Apparently the most trivial example of a collar space is the Hilbert space $H = L^2([-1,1], W)$ whose elements are measurable functions $f: [-1,1] \longrightarrow W$, where W is a separable Hilbert space and

$$\|f\|_H = \left\{ \int_{-1}^{1} \|f(t)\|_W^2 \, dt \right\}^{1/2} < \infty.$$

The action of the algebra $C^\infty([-1,1])$ on H is defined as pointwise multiplication.

Example 1.3. Let M be a smooth complete Riemannian manifold, and let a smooth mapping

$$\chi : M \longrightarrow [-1, 1] \qquad (1.1)$$

be given such that the set $\chi^{-1}(-1, 1)$ is bounded. The Sobolev spaces $H^s(M)$ (defined with the use of powers of the Beltrami–Laplace operator and of the Riemannian volume measure) can naturally be equipped with an action of the algebra $C^\infty([-1,1])$ as follows. Every function $\varphi \in C^\infty([-1,1])$ acts on $H^s(M)$ as the operator of multiplication by the function $\varphi \circ \chi \in C^\infty(M)$. Thus, $H^s(M)$ is a collar space in a natural way.

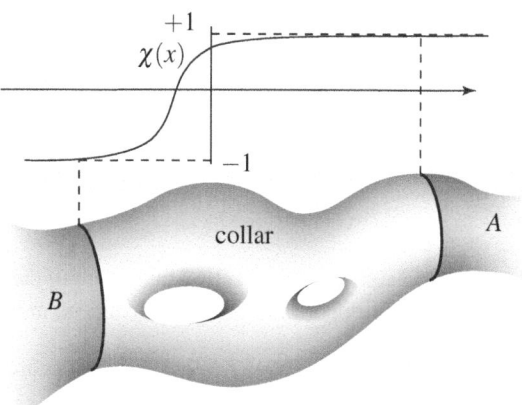

Figure 1.1: A collar space

Let us visualize what happens in this example as well as in more general examples where M is not necessarily a compact C^∞ manifold (see Fig. 1.1). The closed subsets

1.1. Collar Spaces

$A = \chi^{-1}(\{1\})$ and $B = \chi^{-1}(\{-1\})$ play the same role as the sets denoted by the same letters in the classical example given in the Introduction (cf. Fig. 0.1); that is where the surgeries will be applied. The remaining part of M, that is, the set

$$U = M \setminus (A \cup B) = \chi^{-1}\{(-1,1)\}$$

separating A and B, will be referred to as the *collar*. That is where the operators in question should be "local" for the index increments $\triangle_A$ and $\triangle_B$ to be independent indeed.

Of course, the locality that is actually needed to prove superposition theorems for the relative index is much weaker than the usual locality on M. It is not with respect to, say, the algebra $C_0^\infty(M)$ but rather with respect to the action of $C^\infty([-1,1])$, and even for this action, not in the sense that the commutators with the action of the elements of the algebra are compact but in some weaker sense (see the next section). So far, let us define some notions related to locality with respect to the action of $C^\infty([-1,1])$. These are in fact pretty much standard. We start from the notion of support.

Definition 1.4. Let H be a collar space, and let $h \in H$. The *support* of h is the minimal closed subset $K \subset [-1,1]$ such that the following property holds:

$$\varphi \in C^\infty([-1,1]) \text{ and } \operatorname{supp} \varphi \cap K = \varnothing \implies \varphi h = 0.$$

The support of the element h will be denoted by $\operatorname{supp} h$.

Alternatively, one can write

$$\operatorname{supp} h = \bigcap_{\substack{\varphi \in C^\infty([-1,1]) \\ \varphi h = 0}} \varphi^{-1}(\{0\}).$$

This is easily seen to be equivalent to Definition 1.4.

One can readily describe the support of elements of collar spaces in Examples 1.2 and 1.3. In Example 1.2, elements $h \in H$ are (Hilbert space-valued) functions on $[-1,1]$, and $\operatorname{supp} h$ coincides with the support of h in the ordinary sense. In Example 1.3, elements $h \in H^s(M)$ are functions (or distributions) on M, and one has

$$\operatorname{supp} h = \chi(\operatorname{supp}_M h),$$

where $\operatorname{supp}_M h$ is the support of h as a function on M. (Note that $\chi(\operatorname{supp}_M h)$ is necessarily closed.)

The following trivial assertion shows that the notion of support introduced in Definition 1.4 has all natural properties of the ordinary support.

Proposition 1.5. *Let H be a collar space, and let $h, h_1, h_2, \ldots \in H$. Then*

(i) $\operatorname{supp} h$ *is closed.*

(ii) $\operatorname{supp} h = \varnothing$ *if and only if $h = 0$.*

(iii) *One has* $\operatorname{supp}(h_1 + h_2) \subset \operatorname{supp} h_1 \cup \operatorname{supp} h_2$.

(iv) *One has* $\operatorname{supp}(\varphi h) \subset \operatorname{supp}\varphi \cap \operatorname{supp} h$ *for any* $\varphi \in C^\infty([-1,1])$.

(v) *If* $h_n \to h$ *in the space* H *(weakly or strongly), then*

$$\operatorname{supp} h \subset \bigcap_k \overline{\bigcup_{n\geq k} \operatorname{supp} h_n},$$

where the bar stands for closure.

Proof. We omit the proof, which is fairly standard. □

For an arbitrary subset $F \subset [-1,1]$, let $\overset{\circ}{H}_F \subset H$ be the linear manifold of elements $h \in H$ such that $\operatorname{supp} h \subset F$. If $F \subset [-1,1]$ is closed, then $\overset{\circ}{H}_F$ is a closed subspace of H. Indeed, if $h_n \to h$ and $\operatorname{supp} h_n \subset F$, then

$$\operatorname{supp} h \subset \bigcap_k \overline{\bigcup_{n\geq k} \operatorname{supp} h_n} \subset F$$

by Proposition 1.5, (v). For arbitrary $F \subset [-1,1]$, the linear manifold $\overset{\circ}{H}_F$ is not necessarily closed, and we define

$$H_F = \overline{\overset{\circ}{H}_F}.$$

Proposition 1.6. *The space H_F is invariant under the action of $C^\infty([-1,1])$ and hence itself is a collar space.*

Proof is obvious. □

We point out that $H_F \neq H_{\bar F}$ in general. A counterexample can readily be indicated in Example 1.3, where $H^s(M)_{(-1,1)}$ consists of all functions $u \in H^s(M)$ supported in the collar, while $H^s(M)_{[-1,1]}$ coincides with the entire $H^s(M)$.

Finally, we need one technical result.

Proposition 1.7. *Let $F_1,\ldots,F_m \subset [-1,1]$ be subsets such that their closures $\bar F_1,\ldots,\bar F_m$ are pairwise disjoint. Then one has the direct sum decomposition*

$$H_{F_1 \cup \cdots \cup F_m} = H_{F_1} \oplus \cdots \oplus H_{F_m}. \tag{1.2}$$

Proof. Let $V_1 \ldots, V_m$ be disjoint open neighborhoods of $F_1,\ldots,F_m$, respectively, and let $e_1,\ldots,e_m \in C^\infty([-1,1])$ be a partition of unity on $\cup F_j$ subordinate to the cover $\cup V_j \supset \cup F_j$. If $u_j \in H_{F_j}$, $j=1,\ldots,m$, then $\sum u_j \in H_{F_1 \cup \cdots \cup F_m}$. Moreover, if $\sum u_j = 0$, then

$$u_k = e_k \sum_{j=1}^m u_j = 0, \quad k=1,\ldots,m.$$

Finally, if $u \in H_{F_1 \cup \cdots \cup F_m}$, then $e_j u \in F_j$, $j=1,\ldots,m$, and $\sum u_j = u$. □

Remark 1.8. 1. For the direct sum decomposition (1.2) to hold, it does not suffice to require that the sets F_j themselves are pairwise disjoint.

2. The sum on the right-hand side in (1.2) need not be orthogonal.

1.2 Proper Operators and Fredholm Operators

Supports. If an operator A on a function space can be written as an integral operator,

$$[Au](x) = \int A(x,y)u(y)\,dy \tag{1.3}$$

with some (possibly, distributional) kernel $A(x,y)$, then one can speak of the *support* (and *singular support*) of the kernel of A. In fact, a (somewhat weaker) notion of support of the kernel can be defined even if we do not consider the kernel itself. Namely, the support of the kernel can be reconstructed from how the operator transforms the supports of functions on which it acts. This reconstruction can be used as a definition of the support of the kernel even if the kernel itself is not necessarily defined.

Let us use this idea to define the support of an operator $A : H_1 \longrightarrow H_2$ acting between collar spaces. Since the supports of *elements* of collar spaces are defined as subsets of the interval $[-1,1]$, it is natural that the supports of *operators* acting on such spaces will be defined as subsets of the Cartesian product $[-1,1] \times [-1,1]$. First, note that an arbitrary subset $K \subset [-1,1] \times [-1,1]$ can be viewed as the graph of a set-valued mapping, which will be denoted by the same letter,

$$K \colon [-1,1] \longrightarrow 2^{[-1,1]}.$$

Namely, for each $t \in [-1,1]$ we define $K(t)$ to be the set (see Fig. 1.2)

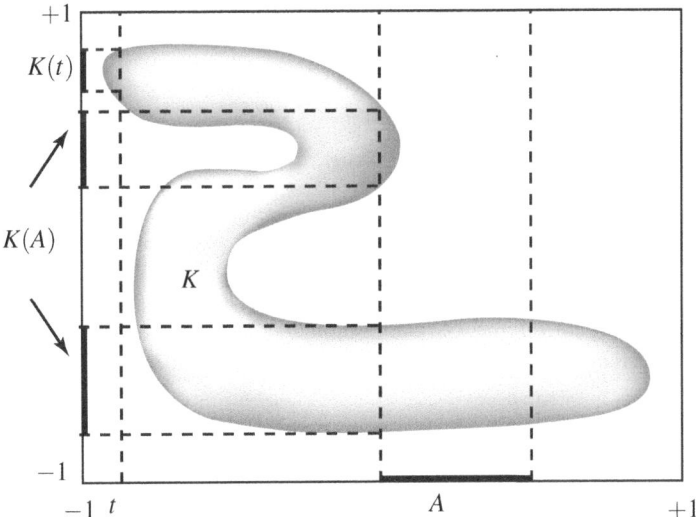

Figure 1.2: A subset $K \subset [-1,1] \times [-1,1]$ as a set-valued mapping

$$K(t) = \{\tau \in [-1,1] : (t,\tau) \in K\}, \tag{1.4}$$

where the images $K(t)$ of a point $t \in [-1,1]$ and $K(A)$ of a subset $A \subset [-1,1]$ are shown.

Now we are in a position to give the desired definition.

Definition 1.9. Let H_1 and H_2 be collar spaces, and let $A: H_1 \longrightarrow H_2$ be a bounded linear operator. One says that *A is supported in a closed subset* $K \subset [-1,1] \times [-1,1]$ if

$$\operatorname{supp} Au \subset K(\operatorname{supp} u) \tag{1.5}$$

for any element $u \in H$. The minimum closed subset $K \subset [-1,1] \times [-1,1]$ with this property is denoted by $\operatorname{supp} A$ and referred to as the *support* of A.

We need yet to prove that there exists a minimum closed subset with this property. The intersection of all such closed sets K is a natural candidate, but the proof is not completely obvious, because in general one only has $(\bigcap K_\alpha)(X) \subset \bigcap K_\alpha(X)$, while the opposite inclusion fails. The desired assertion is valid nevertheless.

Lemma 1.10. *Let $\{K_\alpha\}$ be an arbitrary family of closed subsets of $[-1,1] \times [-1,1]$ satisfying (1.5). Then the intersection $K = \bigcap_\alpha K_\alpha$ satisfies (1.5) as well.*

Proof. Let $u \in H_1$, and let $\tau \notin K(\operatorname{supp} u)$. We claim that $\tau \notin \operatorname{supp} Au$. Indeed, for each α, let U_α be the set of $t \in [-1,1]$ such that $(t,\tau) \notin K_\alpha$. The set U_α is obviously open. Next, $\operatorname{supp} u \subset \bigcup_\alpha U_\alpha$, because it follows from the relation $\tau \notin K(\operatorname{supp} u)$ that for any $t \in \operatorname{supp} u$ one has $(t,\tau) \notin K$ and hence there exists at least one index α such that $(t,\tau) \notin K_\alpha$. Now take a finite subcover of $\operatorname{supp} u$ by U_α, and let $\{e_\alpha\} \subset C^\infty([-1,1])$ be a smooth partition of unity on $\operatorname{supp} u$ subordinate to this subcover. Then

$$Au = \sum_\alpha A(e_\alpha u), \qquad \tau \notin K_\alpha(U_\alpha) \supset \operatorname{supp}(A(e_\alpha u)) \quad \text{for any } \alpha,$$

and hence $\tau \notin \operatorname{supp} Au$ by Proposition 1.5, (iii). □

The supports of operators in collar spaces enjoy many natural properties. For example, the following assertion holds.

Proposition 1.11. *Let H_1, H_2, and H_3 be collar spaces, and let $A: H_1 \longrightarrow H_2$ and $B: H_2 \longrightarrow H_3$ be some operators. Then $\operatorname{supp}(BA) \subset \operatorname{supp} B \circ \operatorname{supp} A$, where the composition on the right-hand side is understood in the sense of set-valued mappings (see Fig. 1.3.)*

Proper operators. The operators for which the index superposition principle will be proved have some special properties related to their supports. Prior to stating the corresponding definition, let us informally explain what these properties are and where they stem from. The operators usually considered in index theory on compact manifolds have the property known as *locality*: their commutators with the operators of multiplication by

1.2. Proper Operators and Fredholm Operators

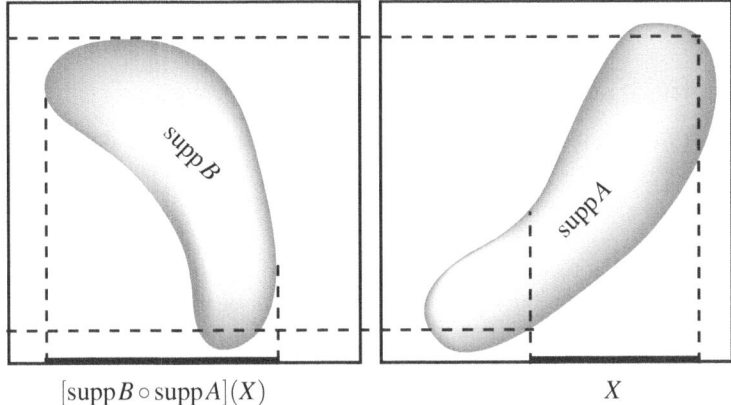

Figure 1.3: Composition of supports

continuous functions are compact. If A is such an operator and $\sum \varphi_j^2(x) = 1$ is a partition of unity on the underlying manifold M, then

$$A = \sum_j \varphi_j \circ A \circ \varphi_j \quad \text{modulo compact operators}.$$

By taking an appropriately fine partition of unity, we can ensure that the kernel of the operator on the right-hand side is supported in an arbitrarily small neighborhood of the diagonal

$$\Delta_M = \{(x,x) : x \in M\} \subset M.$$

This operator has the same index as A.

The locality property itself is not really important for the index superposition theorem in collar spaces to be true. (And it even does not hold, say, in applications to Fourier integral operators.) All we need is that we can make the support of our operator to be localized in an arbitrarily small neighborhood of the diagonal, without changing the index of the operator. To avoid making additional assumptions, we will consider families of operators continuously depending on a parameter $\delta > 0$ whose supports shrink to the diagonal $\Delta = \{(t,t) | t \in [-1,1]\}$ as $\delta \to 0$. If all operators in such a family are Fredholm, then the index is independent of δ. In specific applications, one is usually given a single operator rather than a family; however, as a rule, one can readily include this operator in a family with the shrinking support property.

Let us proceed to a rigorous definition. By Δ_ε we denote the ε-neighborhood of the diagonal Δ, i.e., the set of pairs $(t, \tau) \in [-1,1] \times [-1,1]$ such that $|t - \tau| < \varepsilon$.

Definition 1.12. Let H_1 and H_2 be collar spaces. A family of bounded linear operators

$$A_\delta : H_1 \longrightarrow H_2$$

depending on the parameter $\delta > 0$ is called a *proper operator* if the following conditions are satisfied:

(i) the operator A_δ depends on δ continuously in the operator norm.

(ii) For every $\varepsilon > 0$, there exists a $\delta_0 > 0$ such that, for all $\delta < \delta_0$, one has

$$\operatorname{supp} A_\delta \subset \Delta_\varepsilon.$$

Remark 1.13. One can restate condition (ii) as follows: for $\delta < \delta_0$, the support of $A_\delta u$ is contained in the ε-neighborhood of the support of u for an arbitrary element $u \in H_1$.

Example 1.14. Let us return to Example 1.3, assuming for simplicity that M is a smooth *compact* manifold (which actually does not affect our conclusions). Pseudodifferential operators (ΨDO) on Sobolev spaces on M can be included in operator families satisfying the conditions of Definition 1.12 as follows. Consider an lth-order ΨDO,

$$P \colon H^s(M) \longrightarrow H^{s-l}(M),$$

with kernel $k_P(x,y)$, $x,y \in M$. Take a cutoff function $\varphi_\delta(x,y)$ on $M \times M$ such that φ_δ continuously depends on δ, is equal to 1 in the δ-neighborhood of the diagonal, and is zero outside the 2δ-neighborhood. The ΨDO P_δ with kernel $\varphi_\delta(x,y)k_P(x,y)$ has the same symbol as P and coincides with P provided that δ is sufficiently large. The family P_δ obviously satisfies the desired conditions.

Remark 1.15. The set of proper operators from a collar space H into itself is an algebra. Indeed, it follows from the triangle inequality, Proposition 1.11, and Definition 1.12 that the sum and product of proper operators are again proper operators.

Now we can introduce the class of Fredholm operators to be dealt with in what follows. Recall that an *almost inverse* of a Fredholm operator

$$A \colon H \longrightarrow G$$

is an operator

$$B \colon G \longrightarrow H$$

such that

$$AB = 1 + K_1, \qquad BA = 1 + K_2,$$

where K_1 and K_2 are compact operators on G and H, respectively. An almost inverse of A will be denoted by $A^{[-1]}$; note that the almost inverse is by no means unique, even though the notation suggests otherwise.

Definition 1.16. A *collar Fredholm operator* (a *c-Fredholm operator*) between collar spaces H and G is a proper operator

$$D_\delta \colon H \longrightarrow G$$

with the following properties:

1. The operator is Fredholm for every $\delta > 0$.

2. For every $\delta > 0$, there exists an almost inverse $D_\delta^{[-1]}$ of the operator D_δ such that the family formed by these almost inverses is again a proper operator.

Example 1.17. Let us again return to Example 1.3. If A is an elliptic pseudodifferential operator on a smooth compact manifold M equipped with a smooth function $\chi \colon M \longrightarrow [-1,1]$ and $A^{[-1]}$ is an almost inverse of A, then both operators can be made proper operators in Sobolev spaces on M by the technique used in Example 1.14. Thus, A can be interpreted as a c-Fredholm operator. Note that all elements A_δ of the corresponding family represent the stable homotopy class $[A] \in \mathrm{Ell}(M)$ of the original elliptic operator A; that is why it is expedient not to distinguish between A and the family A_δ unless a misunderstanding is possible. As a rule, we omit the parameter δ in the notation of a proper (in particular, c-Fredholm) operator.

Remark 1.18. The class of c-Fredholm operators is wider than that of Atiyah's general elliptic operators (see Definition 0.6). One relevant example is given by Fourier integral operators (see Section 0.3.4).

1.3 Superposition Principle

In this section, we prove Theorem 0.10. To this end, we need the notion of surgery in collar spaces. Surgery for Atiyah's general elliptic operators was defined in Section 0.2.2 of the Introduction, while surgery in collar spaces was temporarily left to the reader's imagination in Section 0.2.3, and now we describe it in detail.

Surgery of collar spaces and surgery diagrams. Let H and G be collar spaces, let $Q \subset [-1,1]$ be a closed subset, and let $F = [-1,1] \setminus Q$ be the complement of Q. Next, let

$$j \colon H_F \longrightarrow G_F$$

be an isomorphism of collar spaces[1] (not necessarily norm-preserving).

Definition 1.19. The triple (H, G, j) is called a *surgery of collar spaces on the subset Q*. In this situation, we also say that H and G *coincide on* F, or that G is a *modification of H on Q* (and vice versa) or *is obtained from H by surgery on Q*. Furthermore, we write

$$H \stackrel{F}{=} G \quad \text{and} \quad H \stackrel{Q}{\longleftrightarrow} G.$$

This notation omits the isomorphism j for brevity, but one should have in mind that the specific form of the isomorphism is important in the definition: different isomorphisms correspond to different surgeries. The mere existence of an isomorphism is not of much value anyway, because there always exists one provided that H_F and G_F are of the same dimension.

---
[1]Thus, j commutes with the action of $C^\infty([-1,1])$.

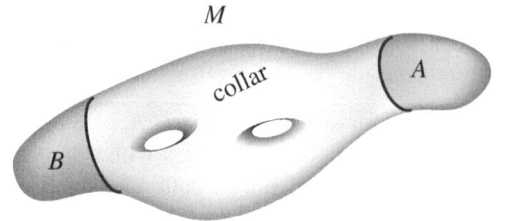

Figure 1.4: The original manifold M

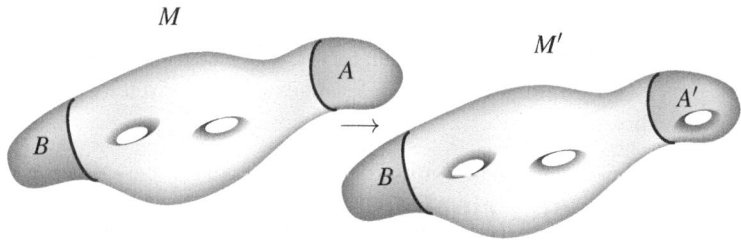

Figure 1.5: Surgery on the set $\{1\}$

Example 1.20. Consider the smooth closed manifold M shown in Fig. 1.4. Let a smooth function $\chi: M \longrightarrow [-1, 1]$ be given, making the Sobolev spaces $H^s(M)$ collar spaces as in Example 1.3. We assume that

$$\chi^{-1}(\{1\}) = A, \qquad \chi^{-1}(\{1\}) = B,$$

so that the collar is the lighter area, just as in Fig. 1.1.

Now let us cut away the part A and paste some different part A', thus obtaining a new smooth closed manifold M' (see Fig. 1.5). The function $\chi|_{M \setminus A}$ extends to be a smooth function on M' if we set $\chi|_{A'} = 1$; thus, the Sobolev spaces $H^s(M')$ are collar spaces as well.

One obviously has

$$H^s(M) \stackrel{[-1,1)}{=} H^s(M'), \qquad \text{or} \qquad H^s(M) \stackrel{\{1\}}{\longleftrightarrow} H^s(M'),$$

the corresponding mapping

$$j: H^s_{[-1,1)}(M) \longrightarrow H^s_{[-1,1)}(M')$$

being defined as follows:

$$[jf](x) = f(x), \quad x \in M \setminus A, \qquad [jf](x) = 0, \quad x \in A', \qquad \text{for } f \in C^\infty_0(M \setminus A).$$

1.3. Superposition Principle

(The mapping j extends to the entire $H^s_{[-1,1)}(M)$ uniquely by continuity.)

In what follows, most often surgeries on the sets $\{1\}$ and $\{-1\}$ are used. To simplify the notation, we omit the braces and write $H \xleftrightarrow{1} G$ rather than $H \xleftrightarrow{\{1\}} G$ and accordingly $H \xleftrightarrow{-1} G$ rather than $H \xleftrightarrow{\{-1\}} G$.

Now we introduce the very important notion of *surgery diagrams*. Let H_i, $i = 1, \ldots, 4$, be collar spaces, and let

$$Q_{12}, Q_{13}, Q_{24}, Q_{34} \in [-1, 1]$$

be closed subsets, whose complements will be denoted by

$$F_{kl} = [-1, 1] \setminus Q_{kl}.$$

Assume that

$$H_1 \xleftrightarrow{Q_{12}} H_2, \quad H_1 \xleftrightarrow{Q_{13}} H_3, \quad H_1 \xleftrightarrow{Q_{23}} H_2, \quad H_1 \xleftrightarrow{Q_{24}} H_2;$$

i.e., H_2 is obtained from H_1 by surgery on Q_{12}, etc., with the underlying isomorphisms

$$j_{12}: H_{1F_{12}} \longrightarrow H_{2F_{12}}, \qquad j_{13}: H_{1F_{13}} \longrightarrow H_{3F_{13}},$$
$$j_{23}: H_{2F_{23}} \longrightarrow H_{3F_{23}}, \qquad j_{24}: H_{2F_{24}} \longrightarrow H_{4F_{24}}.$$

Then we can draw the *surgery diagram*

$$\begin{array}{ccc} H_1 & \xleftrightarrow{Q_{12}} & H_2 \\ {\scriptstyle Q_{13}}\updownarrow & & \updownarrow{\scriptstyle Q_{24}} \\ H_3 & \xleftrightarrow{Q_{34}} & H_4 \end{array}$$

combining all the four surgeries. We say that this surgery diagram *commutes* if the underlying isomorphisms j_{kl} form a commutative diagram

$$\begin{array}{ccc} H_{1F} & \xrightarrow{j_{12}} & H_{2F} \\ {\scriptstyle j_{13}}\downarrow & & \downarrow{\scriptstyle j_{24}} \\ H_{3F} & \xrightarrow{j_{34}} & H_{4F}, \end{array}$$

where

$$F = F_{12} \cap F_{13} \cap F_{24} \cap F_{34} \equiv [-1, 1] \setminus (Q_{12} \cup Q_{13} \cup Q_{24} \cup Q_{34}).$$

(We denote the restrictions of the isomorphisms j_{kl} to subspaces by the same symbol as the homomorphisms themselves.)

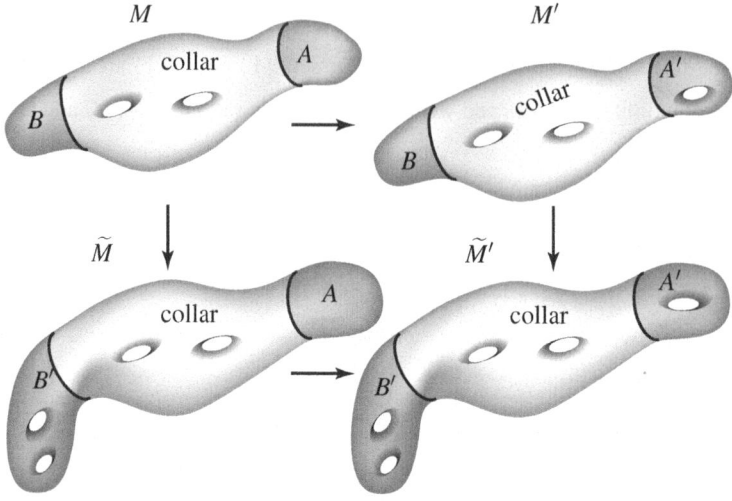

Figure 1.6: Surgery diagram for a manifold M

Example 1.21. Consider the surgery diagram for manifolds shown in Fig. 1.6. With obvious notation, this surgery diagram can be written as

$$\begin{array}{ccc} M & \xleftarrow{A} & M' \\ {\scriptstyle B}\uparrow & & \uparrow{\scriptstyle B} \\ \widetilde{M} & \xleftarrow{A} & \widetilde{M}'. \end{array}$$

If the Sobolev spaces on each of the four manifolds occurring in this diagram are equipped with the structure of collar spaces in the same way as in Example 1.20, then from this surgery diagram for manifolds we obtain the following surgery diagram for Sobolev spaces:

$$\begin{array}{ccc} H^s(M) & \xleftarrow{1} & H^s(M') \\ {\scriptstyle -1}\uparrow & & \uparrow{\scriptstyle -1} \\ H^s(\widetilde{M}) & \xleftarrow{1} & H^s(\widetilde{M}'). \end{array}$$

The latter diagram obviously commutes, because all underlying isomorphisms of Sobolev spaces on the collar are just the identity homomorphism.

Remark. In practical applications, just as in the above example, one usually has $Q_{12} = Q_{34}$ and $Q_{13} = Q_{24}$.

Remark. We have only introduced the simplest kind of surgery diagrams—squares. We will not need more general ones in this book.

1.3. Superposition Principle

Surgery of c-Fredholm operators and surgery diagrams. Now consider the notion of surgery for c-Fredholm operators.

Let H_i and G_i, $i = 1, 2$, be collar spaces, let $Q \subset [-1, 1]$ be a closed subset, and assume that

$$H_1 \xleftrightarrow{Q} H_2, \qquad G_1 \xleftrightarrow{Q} G_2. \tag{1.6}$$

Next, let

$$A_1 \colon H_1 \longrightarrow G_1, \qquad A_2 \colon H_2 \longrightarrow G_2$$

be c-Fredholm operators.

Definition 1.22. One says that A_1 and A_2 are obtained from each other by *surgery on Q* (or *coincide on* $[-1, 1] \setminus Q$) if for any closed subset $K \subset [-1, 1] \setminus Q$ there exists a $\delta_0 > 0$ such that

$$A_1 \delta u = A_2 \delta u \qquad \text{for all } u \in H_1 \text{ with } \operatorname{supp} u \subset K \tag{1.7}$$

for all $\delta < \delta_0$. In this case, we write

$$A_1 \xleftrightarrow{Q} A_2, \quad \text{or} \quad A_1 \xlongequal{F} A_2, \quad F = [-1, 1] \setminus Q.$$

Remark. Of course, there is some abuse of notation in (1.7); more carefully, one should write

$$j_G(A_1 \delta u) = A_2 \delta(j_H u),$$

where j_H and j_G are the isomorphisms underlying the collar space surgeries (1.6). This is well defined. Indeed, $\operatorname{supp} u \subset K \subset [-1, 1] \setminus Q$ and, for sufficiently small δ, one has $\operatorname{supp} A_1 \delta u \subset [-1, 1] \setminus Q$, because A_1 is a proper operator and the distance between the disjoint compact sets K and Q is strictly positive.

Let us return to Example 1.21. Suppose that we have mth-order elliptic differential operators P, P', $\widetilde{P}$, and $\widetilde{P}'$ on M, M', $\widetilde{M}$, and $\widetilde{M}'$, respectively, such that

$$P = P' \text{ on } M \setminus A, \quad P = \widetilde{P} \text{ on } M \setminus B, \quad P' = \widetilde{P}' \text{ on } M' \setminus B, \quad \widetilde{P} = \widetilde{P}' \text{ on } \widetilde{M} \setminus A.$$

Then, treating the operator P as a proper (and c-Fredholm) operator acting in the collar spaces $H^s(M) \longrightarrow H^{s-m}(M)$, etc., we obtain the surgeries

$$P \xleftrightarrow{1} P', \quad P \xleftrightarrow{-1} \widetilde{P}, \quad \widetilde{P} \xleftrightarrow{1} \widetilde{P}', \quad P' \xleftrightarrow{-1} \widetilde{P}'.$$

Let us combine these surgeries into the *surgery diagram*

$$\begin{array}{ccc} P & \xleftrightarrow{1} & P' \\ {\scriptstyle -1}\updownarrow & & \updownarrow{\scriptstyle -1} \\ \widetilde{P} & \xleftrightarrow{1} & \widetilde{P}'. \end{array} \tag{1.8}$$

Such a surgery diagram of c-Fredholm operators is said to *commute* if the two underlying surgery diagrams of collar spaces commute. This is obviously the case in this example, because these underlying diagrams are

$$
\begin{array}{ccc}
H^s(M) \xleftrightarrow{1} H^s(M') & \quad & H^{s-m}(M) \xleftrightarrow{1} H^{s-m}(M') \\
{\scriptstyle -1}\updownarrow \quad\quad \downarrow{\scriptstyle -1} & \quad & {\scriptstyle -1}\updownarrow \quad\quad \updownarrow{\scriptstyle -1} \\
H^s(\widetilde{M}) \xleftrightarrow{1} H^s(\widetilde{M}'), & \quad & H^{s-m}(\widetilde{M}) \xleftrightarrow{1} H^{s-m}(\widetilde{M}'),
\end{array}
$$

and they clearly commute, as was explained above.

Now if P, P', $\widetilde{P}$, and $\widetilde{P}'$ are elliptic *pseudo*differential operators and we only know that

$$\sigma(P) = \sigma(P') \text{ over } M \setminus A, \quad \sigma(P) = \sigma(\widetilde{P}) \text{ over } M \setminus B,$$
$$\sigma(P') = \sigma(\widetilde{P}') \text{ over } M' \setminus B, \quad \sigma(\widetilde{P}) = \sigma(\widetilde{P}') \text{ over } \widetilde{M} \setminus A,$$

where $\sigma(D)$ stands for the principal symbol of a pseudodifferential operator D, then a few more technicalities are needed. First, we act as in Example 1.17 to make these operators c-Fredholm (essentially by modifying them by appropriate families of compact operators). Next, we need to ensure that diagram (1.8) holds—and commutes. To this end, we take a smooth partition of unity

$$e_{-1\delta}^2(t) + e_{0\delta}^2(t) + e_{1\delta}^2(t) = 1$$

on $[-1, 1]$ smoothly depending on the parameter $\delta > 0$ and such that

$$e_{0\delta} = \begin{cases} 1, & |t| \leq 1 - \delta, \\ 0, & |t| \geq 1 - \frac{\delta}{2}, \end{cases} \quad \text{supp}\, e_{-1} \subset \left[-1, -1 + \frac{\delta}{2}\right], \quad \text{supp}\, e_1 \subset \left[1, 1 - \frac{\delta}{2}\right].$$

Then we can replace the operators P, P', $\widetilde{P}$, and $\widetilde{P}'$ by

$$e_{-1\delta} \circ P \circ e_{-1\delta} + e_{0\delta} \circ P \circ e_{0\delta} + e_{1\delta} \circ P \circ e_{1\delta},$$
$$e_{-1\delta} \circ P \circ e_{-1\delta} + e_{0\delta} \circ P \circ e_{0\delta} + e_{1\delta} \circ P' \circ e_{1\delta},$$
$$e_{-1\delta} \circ \widetilde{P} \circ e_{-1\delta} + e_{0\delta} \circ P \circ e_{0\delta} + e_{1\delta} \circ P \circ e_{1\delta},$$
$$e_{-1\delta} \circ \widetilde{P} \circ e_{-1\delta} + e_{0\delta} \circ P \circ e_{0\delta} + e_{1\delta} \circ P' \circ e_{1\delta},$$

respectively. The new operators differ from the original ones by families of compact operators, and it is an easy exercise to check that now we have the commutative diagram (1.8).

Proof of the superposition principle. Now we are in a position to prove Theorem 0.10. First, now that we have introduced c-Fredholm operators and surgery diagrams for such operators, we can restate the theorem without resorting to intuition for understanding any notions. The theorem states that if there is a commutative surgery diagram

1.3. Superposition Principle

$$\begin{array}{ccc} D & \xleftrightarrow{-1} & D_- \\ {}_1\updownarrow & & \updownarrow_1 \\ D_+ & \xleftrightarrow{-1} & D_\pm \end{array} \qquad (1.9)$$

of c-Fredholm operators, then

$$\operatorname{ind}(D) - \operatorname{ind}(D_-) = \operatorname{ind}(D_+) - \operatorname{ind}(D_\pm).$$

To prove Theorem 0.10, we need some lemmas.

Lemma 1.23. *Let $A_1: H_1 \longrightarrow H_2$ and $A_2: H_1 \longrightarrow H_2$ be c-Fredholm operators between collar spaces H_1 and H_2, and suppose that A_1 and A_2 coincide on some open set $F \subset [0,1]$, $A_1 \stackrel{F}{=} A_2$. Then their arbitrary proper almost inverses differ on F by compact operators; more precisely,*

$$A_1^{[-1]} \stackrel{F}{=} A_2^{[-1]} + K,$$

where K is a compact proper operator.

Proof. By the definition of an almost inverse operator, we have

$$A_1 A_1^{[-1]} = 1 + K_1, \qquad A_2^{[-1]} A_2 = 1 + K_2 \qquad (1.10)$$

with some compact proper operators K_1 and K_2. Let us multiply the first relation by $A_2^{[-1]}$ on the left; we obtain

$$A_2^{[-1]} A_1 A_1^{[-1]} = A_2^{[-1]} + A_2^{[-1]} K_1.$$

Next, let us multiply the second relation in (1.10) by $A_1^{[-1]}$ on the right; we obtain

$$A_2^{[-1]} A_1 A_1^{[-1]} \stackrel{F}{=} A_2^{[-1]} A_2 A_1^{[-1]} = A_1^{[-1]} + K_2 A_1^{[-1]}.$$

By combining the last two relations, we see that

$$A_2^{[-1]} \stackrel{F}{=} A_1^{[-1]} + \{K_2 A_1^{[-1]} - A_2^{[-1]} K_1\},$$

which completes the proof, because the expression in braces is a compact operator. $\square$

Lemma 1.24. *Let $A_1: H_1 \longrightarrow H_2$ and $A_2: H_1 \longrightarrow H_2$ be proper operators between collar spaces H_1 and H_2. Suppose that*

$$A_1 \stackrel{F_j}{=} A_2, \qquad j \in \mathcal{J}, \qquad (1.11)$$

for some open sets F_j, $j \in \mathcal{J}$, forming an open cover of the interval $[-1,1]$,

$$\bigcup_{j \in \mathcal{J}} F_j = [-1,1].$$

Then there exists a $\delta_0 > 0$ such that $A_{1\delta} = A_{2\delta}$ for $\delta < \delta_0$. Likewise, if relation (1.11) holds modulo compact proper operators, then $A_{1\delta} = A_{2\delta}$ modulo compact proper operators for $\delta < \delta_0$.

Proof. We can assume without loss in generality that the cover is finite, $\mathscr{J} = \{1,\ldots,N\}$. Let $\{e_j\}_{j=1}^N$ be a smooth partition of unity on $[-1,1]$ subordinate to the cover $\{F_j\}_{j=1}^N$. Then, for each $j = 1,\ldots,N$, $\mathrm{supp}\, e_j$ is a closed subset of F_j, and by Definition 1.22 there exists a $\delta_j > 0$ such that

$$A_{1\delta}u = A_{2\delta}u \qquad \text{for } \mathrm{supp}\, u \subset \mathrm{supp}\, e_j \text{ and } \delta < \delta_j.$$

Set

$$\delta_0 = \min_{j=1,\ldots,N} \delta_j > 0.$$

Then for $\delta < \delta_0$ one has

$$A_{1\delta}u = \sum_{j=1}^N A_{1\delta}(e_j u) = \sum_{j=1}^N A_{2\delta}(e_j u) = A_{2\delta}u.$$

The case of equality modulo compact proper operators can be considered in a completely similar way. □

Remark. Note a useful corollary (which we however do not need in the proof of the theorem): the difference of two arbitrary proper almost inverses of a c-Fredholm operator A is a compact proper operator.

Lemma 1.25. *Assume that all four operators D, D_-, D_+, and $D_\pm$ act in the same pair of spaces. Then Theorem 0.10 is true.*

Proof. Under the assumptions of the lemma, we can represent the difference of the right- and left-hand sides of the desired relation in the form of the index of a product of four operators as follows:

$$\mathrm{ind}\, D - \mathrm{ind}\, D_- - \mathrm{ind}\, D_+ + \mathrm{ind}\, D_\pm = \mathrm{ind}\left(DD_-^{[-1]}D_\pm D_+^{[-1]}\right).$$

However, we have

$$DD_-^{[-1]}D_\pm D_+^{[-1]} \stackrel{(-1,1]}{\equiv} DD^{[-1]}D_+ D_+^{[-1]} \equiv 1$$

(where the symbol $\equiv$ is used to denote equality modulo compact proper operators) and further

$$DD_-^{[-1]}D_\pm D_+^{[-1]} \stackrel{[-1,1)}{\equiv} DD_-^{[-1]}D_- D^{[-1]} \equiv DD^{[-1]} \equiv 1.$$

By Lemma 1.24, we obtain

$$DD_-^{[-1]}D_\pm D_+^{[-1]} \equiv 1,$$

and hence the index of this operator is zero. □

1.3. Superposition Principle

Thus, we have proved Theorem 0.10 for the special case in which all the operators act in the same pair of spaces. Let us proceed to the general case, where the underlying surgery diagrams for collar spaces are nontrivial. Here the key role is played by a special direct sum decomposition that exists for an arbitrary collar space and which is described in the following two technical lemmas.

Lemma 1.26. *Let H be a collar space. Define the following subspaces of H:*

$$H^{(0)} = H_{(-1,1)}, \quad H^{(-)} = H_{[-1,-1/2)} \ominus H_{(-1,-1/2)}, \quad H^{(+)} = H_{(1/2,1]} \ominus H_{(1/2,1)}, \quad (1.12)$$

where $L \ominus L'$ stands for an arbitrary complement (not necessarily the orthogonal complement) of a closed subspace L' of a Hilbert space L. Then one has the direct sum decompositions

$$\begin{aligned} H &= H^{(-)} \oplus H^{(0)} \oplus H^{(+)} = H^{(-)} \oplus H_{(-1,1]} = H_{[-1,1)} \oplus H^{(+)}, \\ H_{[-1,1)} &= H^{(-)} \oplus H^{(0)}, \qquad H_{(-1,1]} = H^{(0)} \oplus H^{(+)} \end{aligned} \quad (1.13)$$

of Hilbert spaces. (These decompositions are in general neither orthogonal, nor do they respect the action of the algebra $C^\infty([-1,1])$.)

Proof. (i) Let us prove that $H = H^{(-)} \oplus H_{(-1,1]}$.

Let $u \in H$. Take a partition of unity $e_1(t) + e_2(t) = 1$ in $C^\infty([-1,1])$ such that $e_1(t) = 0$ for $t > -3/4$ and $e_2(t) = 0$ for $t < -7/8$. Then $u = e_1 u + e_2 u$, where $e_1 u \in H_{[-1,-1/2)}$ and $e_2 u \in H_{(-1,1]}$. Let $e_1 u = v + w$, where $v \in H^{(-)}$ and $w \in H_{(-1,-1/2)}$. Then we finally obtain

$$u = u_1 + u_2, \quad u_1 = v \in H^{(-)}, \quad u_2 = w + e_2 u \in H_{(-1,1]}, \quad (1.14)$$

and moreover,

$$\|u_1\| + \|u_2\| \leq C \|u\|$$

with some constant C.

It remains to prove that the decomposition (1.14) is unique, i.e., $H^{(-)} \cap H_{(-1,1]} = \{0\}$. Let $u \in H^{(-)} \cap H_{(-1,1]}$. Then, in particular, $u \in H_{[-1,-1/2)}$ and $u \in H_{(-1,1]}$. We claim that then $u \in H_{(-1,-1/2)}$ and hence $u = 0$, because $H_{(-1,-1/2)}$ is a complement of $H^{(-)}$. We again write $u = e_1 u + e_2 u$, where e_1 and e_2 are the same as above. Since $u \in H_{[-1,-1/2)}$, it follows that $u = \lim v_n$, where $v_n \in H$ and $\mathrm{supp}\, v_n \subset [-1,-1/2)$; then $e_2 u = \lim e_2 v_n$, $\mathrm{supp}\, e_2 v_n \subset (-1,-1/2)$, and hence $e_2 u \in H_{(-1,-1/2)}$. Now consider $e_1 u$. Since $u \in H_{(-1,1]}$, it follows that $u = \lim w_n$, where $w_n \in H$ and $\mathrm{supp}\, w_n \subset (-1,1]$; then $e_1 u = \lim e_1 w_n$, $\mathrm{supp}\, e_1 w_n \subset (-1,-1/2)$, and hence $e_1 u \in H_{(-1,-1/2)}$.

(ii) In a similar way, one proves that $H_{(-1,1]} = H^{(0)} \oplus H^{(+)}$.

(iii) It follows from (i) and (ii) that $H = H^{(-)} \oplus H^{(0)} \oplus H^{(+)}$.

(iv) The remaining relations in (1.13) can be proved in a similar way. $\square$

Lemma 1.27. *Let us equip the collar space H with the modified action of the algebra $C^\infty([-1,1])$ defined via the original action of this algebra by the formula*

$$\varphi * h = (\varphi \circ \theta) h, \quad (1.15)$$

where $\theta\colon [-1,1] \to [-1,1]$ is a monotone smooth function such that

$$\theta(t) = \begin{cases} -1, & t \in [-1, -1/2], \\ 1, & \theta \in [1/2, 1]. \end{cases}$$

The decompositions (1.13) *respect the action* (1.15); *if we denote the space H equipped with the action* (1.15) *by* $\tilde{H}$, *then*

$$H^{(-)} \subset \tilde{H}_{\{-1\}}, \quad H^{(+)} \subset \tilde{H}_{\{1\}}, \quad \tilde{H}_{[-1,1)} \subset H_{[-1,1)}, \quad \tilde{H}_{(-1,1]} \subset H_{(-1,1]}. \quad (1.16)$$

Proof. The assertion of the lemma becomes clear if we notice that the action (1.15) is the multiplication by $\varphi(-1)$ on $H_{[-1,-1/2)}$ and by $\varphi(1)$ on $H_{(1/2,1]}$. □

Now we can return to the proof of Theorem 0.10. Suppose that the c-Fredholm operators in the theorem are

$$D\colon H \longrightarrow G, \quad D_-\colon H_- \longrightarrow G_-, \quad D_+\colon H_+ \longrightarrow G_+, \quad D_\pm\colon H_\pm \longrightarrow G_\pm.$$

The collar spaces in which these operators act form the commutative surgery diagrams

$$\begin{array}{ccccccc} H & \xleftarrow{-1} & H_- & & G & \xleftarrow{-1} & G_- \\ {\scriptstyle 1}\uparrow & & \downarrow{\scriptstyle 1} & & {\scriptstyle 1}\uparrow & & \downarrow{\scriptstyle 1} \\ H_+ & \xleftarrow{-1} & H_\pm, & & G_+ & \xleftarrow{-1} & G_\pm. \end{array} \quad (1.17)$$

Consider, for example, the first of these diagrams. We have the underlying isomorphisms

$$\begin{aligned} j_-\colon H_{(-1,1]} &\longrightarrow H_{-(-1,1]}, & j_+\colon H_{[-1,1)} &\longrightarrow H_{+[-1,1)}, \\ j_\mp\colon H_{+(-1,1]} &\longrightarrow H_{\pm(-1,1]}, & j_\pm\colon H_{-[-1,1)} &\longrightarrow H_{\pm[-1,1)}, \end{aligned} \quad (1.18)$$

which form the commutative surgery diagram

$$\begin{array}{ccc} H_{(-1,1)} & \xrightarrow{j_-} & H_{-(-1,1)} \\ {\scriptstyle j_+}\downarrow & & \downarrow{\scriptstyle j_\pm} \\ H_{+(-1,1)} & \xrightarrow{j_\mp} & H_{\pm(-1,1)} \end{array} \quad (1.19)$$

of collar spaces.

Lemma 1.28. *The isomorphisms* (1.18) *can be extended to isomorphisms*

$$\begin{aligned} \tilde{j}_-\colon H &\longrightarrow H_-, & \tilde{j}_+\colon H &\longrightarrow H_+, \\ \tilde{j}_\mp\colon H_+ &\longrightarrow H_\pm, & \tilde{j}_\pm\colon H_- &\longrightarrow H_\pm \end{aligned} \quad (1.20)$$

1.3. Superposition Principle

of Hilbert spaces such that the diagram

$$\begin{array}{ccc} H & \xrightarrow{\tilde{j}_-} & H_- \\ \tilde{j}_+ \downarrow & & \downarrow \tilde{j}_\pm \\ H_+ & \xrightarrow{\tilde{j}_\mp} & H_\pm \end{array} \qquad (1.21)$$

commutes and the mappings (1.20) *themselves commute with the action* (1.15) *of the algebra* $C^\infty([-1,1])$.

Proof. We use Lemma 1.26 and seek the desired diagram (1.21) in the form

$$\begin{array}{ccc} H^{(-)} \oplus H^{(0)} \oplus H^{(+)} & \xrightarrow{\tilde{j}_-} & H_-^{(-)} \oplus H_-^{(0)} \oplus H_-^{(+)} \\ \tilde{j}_+ \downarrow & & \downarrow \tilde{j}_\pm \\ H_+^{(-)} \oplus H_+^{(0)} \oplus H_+^{(+)} & \xrightarrow{\tilde{j}_\mp} & H_\pm^{(-)} \oplus H_\pm^{(0)} \oplus H_\pm^{(+)}, \end{array} \qquad (1.22)$$

where at the vertices we have used the decompositions (1.13) and all four isomorphisms in the diagram should respect the direct sum structure (i.e., be block diagonal). (To achieve this, we use the freedom in the choice of subspaces with $(-)$ and $(+)$ subscripts.) Thus, we should actually construct three separate commutative diagrams,

$$\begin{array}{ccc|ccc|ccc} H^{(0)} & \xrightarrow{j_-} & H_-^{(0)} & H^{(-)} & \xrightarrow{\tilde{j}_-} & H_-^{(-)} & H^{(+)} & \xrightarrow{j_-} & H_-^{(+)} \\ j_+ \downarrow & & \downarrow j_\pm & \tilde{j}_+ \downarrow & & \downarrow \tilde{j}_\pm & \tilde{j}_+ \downarrow & & \downarrow \tilde{j}_\pm \\ H_+^{(0)} & \xrightarrow{j_\mp} & H_\pm^{(0)}, & H_+^{(-)} & \xrightarrow{\tilde{j}_\mp} & H_\pm^{(-)}, & H_+^{(+)} & \xrightarrow{j_\mp} & H_\pm^{(+)}, \end{array} \qquad (1.23)$$

where the tilde is placed over yet unknown mappings (see below). The first of these diagrams is already known; it just expresses the fact that the first surgery diagram in (1.17) commutes.

Let us construct the second diagram in (1.23). We can assume without loss of generality that all spaces in it are infinite-dimensional. (If this is not the case, then we can add an infinite-dimensional Hilbert space (say, l^2) as a direct summand to these spaces and the identity operators on l^2 to the original operators.) We choose the complements $H^{(-)}$ and $H_-^{(-)}$ (see the statement of Lemma 1.26) arbitrarily. The vertical mappings in this diagram are known (because

$$j_+ : H_{[-1,1)} \equiv H^{(-)} \oplus H^{(0)} \longrightarrow H_{+[-1,1)} \equiv H_+^{(-)} \oplus H_+^{(0)}$$

is given, and likewise for $j_\pm$), and this uniquely determines the spaces

$$H_+^{(-)} = j_+(H^{(-)}), \qquad H_\pm^{(-)} = j_\pm(H_-^{(-)}).$$

(One can readily verify that they are complements of the appropriate subspaces as required in Lemma 1.26.) It remains to determine the horizontal arrows $\widetilde{j}_-$ and $\widetilde{j}_\mp$. For the first arrow, one can take an arbitrary isomorphism (which always exists between separable Hilbert spaces), and then one should take

$$\widetilde{j}_\mp = j_\pm \circ \widetilde{j}_- \circ j_+^{-1}$$

to make the diagram commute.

The construction of the third diagram in (1.23) is completely similar, with vertical and horizontal arrows exchanging their roles.

The resulting isomorphisms commute with the action (1.15), which follows from the first two embeddings in (1.16).

The proof of the lemma is complete. □

Now we see that the four collar spaces (with respect to the action (1.15)) in the first surgery diagram in (1.17) are isomorphic and hence can be identified with each other. The same is true for the collar spaces in the second diagram in (1.17). The operators D, D_-, D_+, and $D_\pm$ remain c-Fredholm with respect to the new action of $C^\infty([-1,1])$; they all act in the same pair of spaces and satisfy the surgery diagram (1.9) in the sense of the new action. (The last assertion follows from the third and fourth embeddings in (1.16).) Thus, we have reduced the analysis to the case covered by Lemma 1.25. The proof of Theorem 0.10 is complete. □

Chapter 2

Superposition Principle for K-Homology

2.1 Preliminaries

In Chapter 1, we proved the superposition principle for the relative index in the fairly broad setting of collar spaces. However, although the index is possibly the most important homotopy invariant of elliptic operators, it is by no means the only one, and it is natural to ask whether the principle can be generalized to cover other invariants as well. This and the next chapter deal exactly with this topic. The results of the present chapter were obtained in [63].

Let us start by very informally discussing a natural framework to be used for such a generalization. A *stable homotopy invariant* of elliptic operators is by definition a function on the group Ell of stable homotopy classes of elliptic operators. Thus, if we plan to establish a superposition principle valid for "arbitrary" additive stable homotopy invariants, it is apparently inevitable that we end up dealing with Ell and stating our theorems in terms of Ell (directly or indirectly). Note that the group Ell has been computed in many cases. For example, the stable homotopy classification of elliptic pseudodifferential operators on a smooth closed manifold M is given by the formula (e.g., see [11, 19, 35])

$$\text{Ell}(M) = K_0(M), \qquad (2.1)$$

where $K_0(M)$ is the (even) K-homology group of the manifold M. The same formula is true for (appropriately defined) elliptic operators on manifolds M with conical singularities or edges and, more generally, on stratified manifolds (see [49, 50, 75]). The reader should not get the impression that (2.1) always holds. A counterexample is given by elliptic operators on a manifold M with corners; in the simplest case, one has

$$\text{Ell}(M) = K_0(M^\#), \qquad (2.2)$$

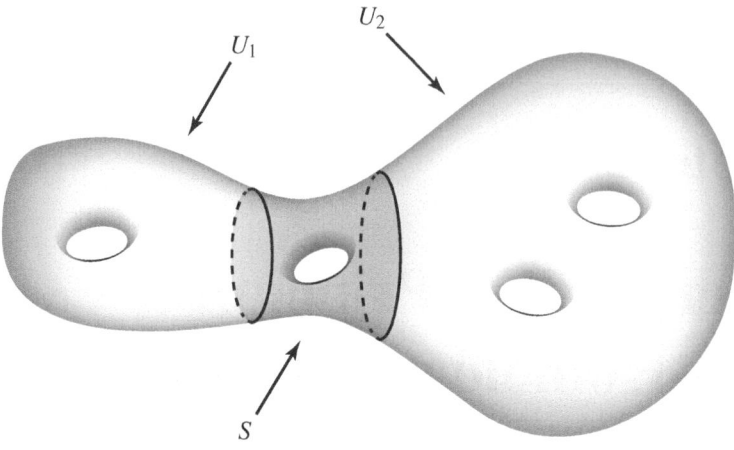

Figure 2.1: A smooth manifold M with a collar S

where $M^\#$ is the "dual" stratified manifold (see [51,52]). Anyway, in all of these examples the group $\mathrm{Ell}(M)$ is isomorphic to some K-homology group, and since the latter is a much more convenient object to deal with than the former, it is reasonable to adopt the point of view that a "general" superposition principle theorem should be stated as an equality in a K-homology group. Moreover, in the spirit of noncommutative geometry [26], we consider the K-homology groups of an arbitrary (not necessarily commutative) C^*-algebra A. [This of course includes the above examples if one sets $A = C(M)$ (the C^*-algebra of continuous functions on M) or $A = C(M^\#)$, because $K_0(M) = K^0(C(M))$.]

This having been settled, next, just as in Chapter 1, we should take care of two things. First, our operators should be *local* in some sense. What is the exact meaning of the word? Second, since the superposition principle basically says that the effect of two simultaneous independent modifications of an elliptic operator on its homotopy invariants is the sum of effects that would be caused by each of the modifications separately, we need to define what *independent modifications* are. In other words, what plays the role of separated regions of a manifold in our noncommutative C^*-algebra setting?

Fortunately, both questions can be answered in a natural way. First, the elements of K-homology groups of a C^*-algebra A correspond to Hilbert space operators D whose commutators with all operators of a representation of A are compact. It is such operators D that are said to be *local*. To address the second question, let us cast a look on the already familiar case of smooth closed manifolds. Let D be an elliptic pseudodifferential operator of order zero on the manifold M shown in Fig. 2.1. (The operator is assumed to act in

2.1. Preliminaries

L^2-spaces of sections of some bundles on M.) This operator is local, which means that

The commutator $[D, \varphi]$ is a compact operator for any function $\varphi \in C(M)$.

Suppose that we wish to modify D on U_1, thus obtaining a new elliptic operator D_1. A modification on U_1 means precisely that the operator "remains the same" on $S \cup U_2$ (or is possibly altered by a compact operator, which does not affect any homotopy invariants); this property can be expressed as follows:

$$\varphi D \psi \equiv \varphi D_1 \psi \mod (\text{compact operators}) \text{ for any } \varphi, \psi \in C(M) \text{ supported in } S \cup U_2.$$

The set of functions $\varphi \in C(M)$ supported in $S \cup U_2$ is an ideal in $C(M)$; let us denote it by J_2. Then our property becomes

$$\varphi D \psi \sim \varphi D_1 \psi \qquad \forall \varphi, \psi \in J_2, \tag{2.3}$$

where we write $B \sim C$ if the difference $B - C$ is a compact operator, and we will simply say that D and D_1 agree on J_2. Now if we modify D on U_2, thus obtaining a new elliptic operator D_2, then D and D_2 should coincide on $S \cup U_1$, which can be expressed as

$$\varphi D \psi \sim \varphi D_2 \psi \qquad \forall \varphi, \psi \in J_1, \tag{2.4}$$

where $J_1 \subset C(M)$ is the ideal of functions supported in $S \cup U_1$. The ideals J_1 and J_2 have the important property that

$$J_1 + J_2 = C(M); \tag{2.5}$$

i.e., every function in $C(M)$ can be represented as the sum of two functions, one in J_1 and one in J_2. The intersection $J = J_1 \cap J_2$ is just the algebra $C_0(S)$ of compactly supported functions on the (open) collar S.

Now we can state the simplest superposition principle for elliptic operators on M as follows. Suppose that D, D_1, D_2, and D_{12} are four elliptic operators on M such that the following diagram of modifications holds:

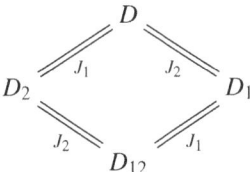

(Here we write, say, $D_2 \underset{J_1}{=\!=} D$ to indicate that D_2 and D agree on J_1.) Then

$$[D_2] - [D] = [D_{12}] - [D_1], \tag{2.6}$$

where $[B] \in K^0(C(M)) = K_0(M)$ is the K-homology class corresponding to an elliptic operator B.

Note that this statement is virtually ready for being translated into the language of noncommutative geometry. It suffices to replace $C(M)$ by an arbitrary C^*-algebra A containing two ideals J_1 and J_2 such that $A = J_1 + J_2$ and, instead of elliptic operators on M, consider Fredholm operators commuting modulo compact operators with a representation of A. However, recall that what was said above is only a very informal scheme in that we have omitted many important details. Precise statements will be given below, and for now let us only make a couple of revealing remarks.

- The correspondence between elliptic operators and K-homology classes, especially in the general case, is not that straightforward and involves a *normalization procedure* (which in particular ensures that the operator is bounded as a mapping of a certain Hilbert space into itself).
- Modification of operators often means that the original operator D and the modified operator D_1 act in distinct spaces, and so one should give a precise definition of what a comparison like (2.3) means.

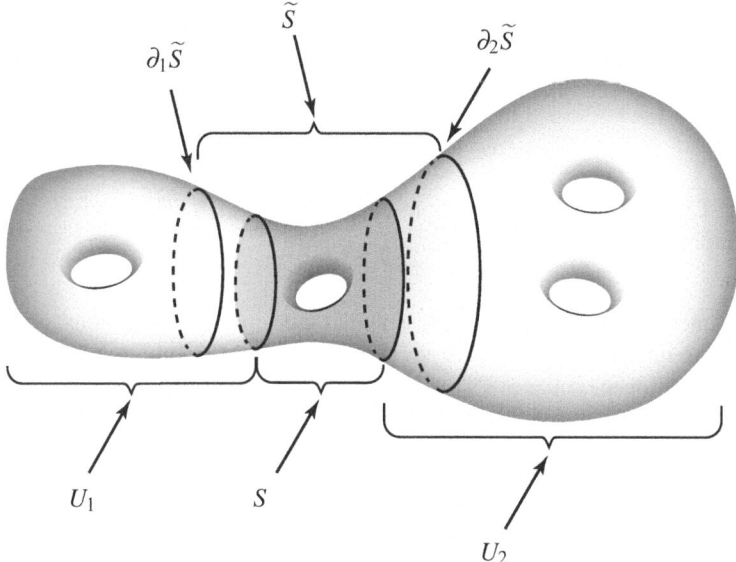

Figure 2.2: The manifold M where the original operator D lives

- Moreover, in fact, the most interesting superposition theorems involve modifications not only of operators and spaces where they act but also of the underlying manifolds (or, in the noncommutative analysis setting, of the underlying C^*-algebras). That is, each of the four operators D, D_1, D_2, and D_{12} occurring in such a theorem might generate a K-homology class of its own C^*-algebra. While it is quite possible to prove superposition theorems in this situation, we trade generality for clarity and adopt a coarser approach in which everything is considered over a C^*-algebra common for all four operators. This can

2.1. Preliminaries

be best illustrated by Fig. 2.2 (which pertains to the commutative case).

This figure shows the manifold M on which the original operator D is defined. Suppose that the independent modifications of this operator on U_1 and U_2 involve changing M on $U_1 \setminus \widetilde{S}$ and $U_1 \setminus \widetilde{S}$, respectively, where the closed domain $\widetilde{S} \subset M$ contains the collar S but does not necessarily coincide with its closure $\overline{S}$. Thus, all four manifolds to be involved in the superposition theorem (the original manifold M, the manifolds M_1 and M_2 obtained from it by changes on $U_1 \setminus \widetilde{S}$ and $U_1 \setminus \widetilde{S}$, respectively, and the manifold M_{12} obtained by the simultaneous change on $U_1 \setminus \widetilde{S}$ and $U_1 \setminus \widetilde{S}$) have the common subset $\widetilde{S}$. Let $\partial_1 \widetilde{S}$ and $\partial_2 \widetilde{S}$ be the components of the boundary $\partial \widetilde{S}$ lying in U_1 and U_2, respectively. The algebra

$$A = \{f \in C(\widetilde{S}) : f|_{\partial_1 \widetilde{S}} = \text{const}, \quad f|_{\partial_2 \widetilde{S}} = \text{const}\}$$

can be interpreted as a subalgebra of $C(M)$: we just extend every $f \in A$ as a continuous function constant on $U_1 \setminus \widetilde{S}$ and on $U_1 \setminus \widetilde{S}$. The ideals $\widetilde{J}_j = J_j \cap A$ of the algebra A still have the property

$$\widetilde{J}_1 + \widetilde{J}_2 = A.$$

Next, the embedding $A \subset C(M)$ gives rise to a mapping $K^0(C(M)) \longrightarrow K^0(A)$ in K-homology, and thus every elliptic operator on M defines a class in $K^0(A)$. The same is true, with the same algebra A and ideals $\widetilde{J}_1$ and $\widetilde{J}_2$, for elliptic operators on the other three manifolds, and the superposition theorem can be stated in the K-homology of A. The passage to the narrower algebra A does result in a loss of some information, but this is the price we have to pay for avoiding awkwardness.

• Finally, note that, although the material of this and the next chapters is closely related to that in Chapter 1, the results of Chapter 1 *do not follow* (at least, directly) from the ones given here. The reasons are twofold. First, we abandoned the Fréchet algebra $C^\infty[-1,1]$ for the much more robust setting of C^*-algebras and hence actually restricted ourselves to operators of order zero in L^2-type spaces. This is no big deal, because order reduction takes care of that. Second, which is more important, c-Fredholm operators may be nonlocal with respect to the action of the algebra $C^\infty[-1,1]$, while in the framework of K-homology we work only with local operators. Thus, we cannot apply the results, say, to Fourier integral operators. That said, for local operators everything goes pretty well. If the algebra A is unital, then one can readily obtain a relative index theorem from a superposition theorem like (2.6) in K-homology, because the embedding $\mathbb{C} \subset A$ of C^*-algebras (where $z \mapsto z \cdot 1$) induces the index homomorphism $\text{ind} \colon K^0(A) \longrightarrow K^0(\mathbb{C}) \equiv \mathbb{Z}$.

Now let us proceed to precise, rigorous statements.

2.2 Fredholm Modules and K-Homology

Here, mainly following [35, Chapter 8], we recall how the K-homology groups of a C^*-algebra A are defined.

Some notation and conventions. We refer the reader to [27, 66] for basic information concerning C^*-algebras and their representations. By an ideal in a C^*-algebra we always mean a two-sided closed $*$-ideal. A homomorphism of C^*-algebras is always understood as a $*$-homomorphism and is assumed to be unital (unless explicitly specified otherwise) whenever both algebras are unital.

Let H be a Hilbert space. We denote the C^*-algebra of bounded operators on H by $\mathfrak{B}(H)$ and the ideal of compact operators by $\mathfrak{K}(H)$. Likewise, if H_1 and H_2 are two Hilbert spaces, then $\mathfrak{B}(H_1, H_2)$ stands for the Banach space of bounded linear operators from H_1 to H_2 and the symbol $\mathfrak{K}(H_1, H_2)$ is used to denote the subspace of compact operators. For $B, C \in \mathfrak{B}(H)$ (or $\mathfrak{B}(H_1, H_2)$), we write $B \sim C$ if $B - C \in \mathfrak{K}(H)$ (resp., $\mathfrak{K}(H_1, H_2)$).

Let $\rho : A \longrightarrow \mathfrak{B}(H)$ be a representation of a C^*-algebra A on a Hilbert space H. By $\approx$ we denote equality modulo locally compact operators; i.e.,

$$B, C \in \mathfrak{B}(H), \; B \approx C \iff \rho(\varphi)(B-C), \; (B-C)\rho(\varphi) \in \mathfrak{K}(H) \; \forall \varphi \in A.$$

Likewise, if $\rho_1 : A \longrightarrow \mathfrak{B}(H_1)$ and $\rho_2 : A \longrightarrow \mathfrak{B}(H_2)$ are representations of A on two Hilbert spaces, then for $B, C \in \mathfrak{B}(H_1, H_2)$ we write $B \approx C$ if

$$\rho_2(\varphi)(B-C), \; (B-C)\rho_1(\varphi) \in \mathfrak{K}(H_1, H_2) \; \forall \varphi \in A.$$

Clearly, relations $\approx$ and $\sim$ are the same if the algebra A is unital.

Fredholm modules. Let A be a C^*-algebra, not necessarily unital. (We will however assume that it is separable.)

Definition 2.1. An *(ungraded) Fredholm module* over A is a triple $x = (H, \rho, F)$, where H is a Hilbert space, $\rho : A \longrightarrow \mathfrak{B}(H)$ is a representation of A on H, and $F \in \mathfrak{B}(H)$ is an operator such that

$$[F, \rho(\varphi)] \sim 0 \quad \forall \varphi \in A \quad (locality), \qquad F \approx F^*, \qquad F^2 \approx 1. \tag{2.7}$$

A *graded Fredholm module* is defined in the same way with the additional requirement that the space H is $\mathbb{Z}_2$-graded, that is, is represented as the direct sum

$$H = H_+ \oplus H_-, \tag{2.8}$$

the representation ρ is even, that is, preserves the grading,

$$\rho(A)H_+ \subset H_+, \qquad \rho(A)H_- \subset H_-,$$

and the operator F is odd, that is, satisfies

$$FH_+ \subset H_-, \qquad FH_- \subset H_+.$$

2.2. Fredholm Modules and K-Homology

In the graded case, if we represent all operators by 2×2 block matrices corresponding to the decomposition (2.8), then we have

$$\rho(\varphi) = \begin{pmatrix} \rho_+(\varphi) & 0 \\ 0 & \rho_-(\varphi) \end{pmatrix}, \quad \varphi \in A; \qquad F = \begin{pmatrix} 0 & W \\ V & 0 \end{pmatrix},$$

and moreover,

$$WV \approx 1, \quad VW \approx 1, \quad W \approx V^*, \quad \rho_+(\varphi)W \sim W\rho_-(\varphi), \quad \rho_-(\varphi)V \sim V\rho_+(\varphi) \,\forall \varphi \in A.$$

A Fredholm module $x = (H, \rho, F)$ is said to be *degenerate* if all relations in (2.7) are satisfied exactly rather than modulo (locally) compact operators.

A Fredholm module $x = (H, \rho, F)$ is said to be *nondegenerate* if the (necessarily closed) subspace $\rho(A)H \subset H$ coincides with the entire H. (Strangely enough, being nondegenerate is *not* the opposite of being degenerate.)

Two Fredholm modules $x = (H, \rho, F)$ and $x' = (H', \rho', F')$ are said to be *unitarily equivalent* if there exists a unitary operator $U : H \longrightarrow H'$ such that U intertwines the representations ρ and ρ' (that is, $U\rho(\varphi) = \rho'(\varphi)U$ for every $\varphi \in A$) and $F' = UFU^{-1}$.

Finally, two Fredholm modules (H, ρ, F_0) and (H, ρ, F_1) corresponding to one and the same representation $\rho : A \longrightarrow \mathfrak{B}(H)$ are said to be *homotopic* if they can be included in a family $x_t = (\rho, H, F_t), t \in [0, 1]$, of Fredholm modules such that the operator function $t \mapsto F_t$ is norm continuous.

In the graded case, the operator U in the definition of unitary equivalence is required to preserve the grading, and the family x_t in the definition of homotopic Fredholm modules should consist of graded Fredholm modules.

K-homology. Two ungraded Fredholm modules x and x' are said to be *equivalent* if there exists a degenerate module x'' such that the modules $x \oplus x''$ and $x' \oplus x''$ are unitarily equivalent to homotopic Fredholm modules.

Proposition 2.2. *The operation of direct sum of modules gives rise to a well-defined operation on the set $K^1(A)$ of equivalence classes of ungraded Fredholm modules. This operation makes $K^1(A)$ an abelian group, the neutral element being the equivalence class of (each and every) degenerate module.*

The group $K^1(A)$ is called the *odd K-homology group* of the algebra A. The definition of the *even K-homology group* $K^0(A)$ is completely similar; the only difference is that one considers graded Fredholm modules.

2.3 Superposition Principle

Throughout this section, A is a C^*-algebra, possibly nonunital, and $J_1, J_2 \subset A$ are two ideals such that
$$J_1 + J_2 = A.$$
(The sum is not assumed to be direct; i.e., the intersection $J_0 = J_1 \cap J_2$ may be nonempty and is so in cases of interest.)

The results stated below hold for $K^0(A)$ as well as $K^1(A)$, and in what follows we write $K^*(A)$ instead of both, tacitly assuming that all Fredholm modules considered are graded or not depending on whether $* = 0$ or $* = 1$, respectively.

Block matrix form of a Fredholm module. Let $x = (H, \rho, F)$ be a nondegenerate Fredholm module over A. We use the ideals J_1, J_2, and J_0 to represent the operator F in the form of a 3×3 block matrix.

To this end, recall that if $J \subset A$ is an ideal, then the set $H_J = \rho[J]H \subset H$ of vectors of the form $\rho(\varphi)\xi$, $\varphi \in J$, $\xi \in H$, is a closed subspace of H (e.g., see [35, Exercise 1.9.17]), and the orthogonal projection P_J onto this subspace in H commutes with the operators $\rho(\varphi)$ for every $\varphi \in A$. Consequently, H_J is an A-invariant subspace, and one has a well-defined representation
$$\rho \upharpoonright H_J \colon A \longrightarrow \mathfrak{B}(H_J)$$
(which we denote again by ρ if no confusion can arise).

Let us apply this to our ideals J_1, J_2, and $J_0 = J_1 \cap J_2$. Thus, we have three invariant subspaces of H,
$$H_{J_1} = \rho[J_1]H, \quad H_{J_2} = \rho[J_2]H, \quad H_{J_0} = \rho[J_0]H.$$

Lemma 2.3. *One has*
$$H_{J_1} + H_{J_2} = H, \quad H_{J_1} \cap H_{J_2} = H_{J_0}.$$

Proof. The first relation readily follows from the nondegeneracy assumption. The second relation is proved in greater generality (for representations on Hilbert modules) in Lemma 3.5 in Chapter 3. $\square$

Set
$$H_0 = H_{J_0}, \quad H_1 = H_{J_1} \ominus H_0, \quad H_2 = H_{J_2} \ominus H_0.$$
(The symbol $\ominus$ is used to denote the operation of taking the orthogonal complement.)

Lemma 2.4. *One has the decomposition*
$$H = H_1 \oplus H_0 \oplus H_2 \tag{2.9}$$
of H into an orthogonal sum of A-invariant subspaces.

2.3. Superposition Principle

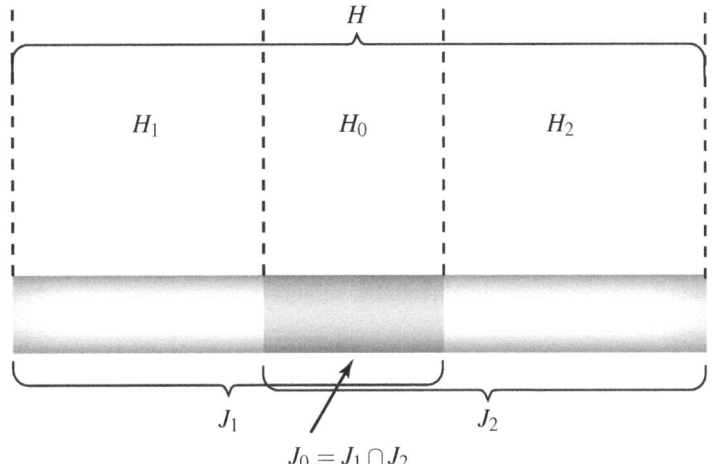

Figure 2.3: Ideals in the algebra A and the direct sum decomposition of H

Proof. In view of Lemma 2.3, the only thing to prove is that $H_1 \perp H_2$. Let $\xi_j \in H_j$, $j = 1, 2$. Let $\{u_\alpha\}$ be the approximate unit for J_1. Then $\rho(u_\alpha)\xi_1 \to \xi_1$ and $\rho(u_\alpha)\xi_2 = 0$, since we simultaneously have $\rho(u_\alpha)\xi \in H_2$ and $\rho(u_\alpha)\xi \in H_{J_1} = H_1 \oplus H_0$. Thus, for the inner product $\langle \xi_1, \xi_2 \rangle$ we obtain

$$\langle \xi_1, \xi_2 \rangle = \lim \langle \rho(u_\alpha)\xi_1, \xi_2 \rangle = \lim \langle \xi_1, \rho(u_\alpha)\xi_2 \rangle = 0. \qquad \square$$

We denote the orthogonal projections onto H_j by P_j, $j = 0, 1, 2$. The direct sum decomposition (2.9) (mind the order of the direct summands—we place H_0 in the middle) and its relationship with the ideals J_1, J_2, and J_0 is visualized in Fig. 2.3.

From now on, for arbitrary $\varphi \in A$ we often write simply φ instead of $\rho(\varphi)$, just to make formulas shorter. We occasionally do so even if there are several representations involved; which one is meant is always clear from the context. Furthermore, given $\varphi \in A$, by $\varphi_1 \in J_1$ and $\varphi_2 \in J_2$ we denote arbitrary elements such that $\varphi = \varphi_1 + \varphi_2$.

Let us represent F as a 3×3 block matrix in the decomposition (2.9). Without loss of generality, we assume that F is self-adjoint. Next, note that

$$\varphi P_1 F P_2 = \varphi_1 P_1 F P_2 = P_1 \varphi_1 F P_2 \sim P_1 F \varphi_1 P_2 = 0 \qquad \forall \varphi \in A;$$

i.e., $P_1 F P_2 \approx 0$ and likewise $P_2 F P_1 \approx 0$.

Hence

$$F \approx \begin{pmatrix} a & b & 0 \\ b^* & c & d \\ 0 & d^* & e \end{pmatrix}, \qquad a = a^* \quad c = c^*, \quad e = e^*, \tag{2.10}$$

where all entries are local. The condition $F^2 \approx 1$ can be rewritten in the form of the equations[1]

$$a^2 + bb^* \approx 1, \quad ab + bc \approx 0, \quad cd + de \approx 0, \quad d^*d + e^2 \approx 1, \quad bd \approx 0, \qquad (2.11)$$

$$\varphi b^*b \sim \varphi_1(1-c^2), \quad \varphi dd^* \sim \varphi_2(1-c^2) \qquad \forall \varphi \in A. \qquad (2.12)$$

Moreover, the last condition in (2.11) is satisfied automatically, because

$$\varphi bd = (\varphi_1 b)d \sim (b\varphi_1)d = b(\varphi_1 d) = 0,$$

and condition (2.12) can be obtained from the relation

$$b^*b + dd^* + c^2 \approx 1.$$

Fredholm modules agreeing on an ideal. Let $x = (H, \rho, F)$ and $\widetilde{x} = (\widetilde{H}, \widetilde{\rho}, \widetilde{F})$ be Fredholm modules over A, and let $J \subset A$ be an ideal. Just as above, we set $H_J = \rho[J]H$ and $\widetilde{H}_J = \widetilde{\rho}[J]\widetilde{H}$ and denote by P_J and $\widetilde{P}_J$ the orthogonal projections onto H_J in H and onto $\widetilde{H}_J$ in $\widetilde{H}$, respectively.

Definition 2.5. We say that x and $\widetilde{x}$ *agree on the ideal J* if there exists a unitary operator $T: H_J \to \widetilde{H}_J$ intertwining the representations $\rho \upharpoonright H_J$ and $\widetilde{\rho} \upharpoonright \widetilde{H}_J$, preserving the grading in the case of graded modules, and satisfying the relation

$$TP_J F P_J \approx \widetilde{P}_J \widetilde{F} \widetilde{P}_J T.$$

In this case, we write $x \xrightarrow[J]{T} \widetilde{x}$.

Recall that the term "intertwining" means that

$$T(\rho \upharpoonright H_J)(\varphi) = (\widetilde{\rho} \upharpoonright \widetilde{H}_J)(\varphi) T \qquad \forall \varphi \in A.$$

The operator T being unitary, Definition 2.5 means in particular that the representations $\rho \upharpoonright H_J$ and $\widetilde{\rho} \upharpoonright \widetilde{H}_J$ are equivalent.

Cutting and pasting. Let $x = (H, \rho, F)$ and $\widetilde{x} = (\widetilde{H}, \widetilde{\rho}, \widetilde{F})$ be nondegenerate Fredholm modules over A that agree on the ideal $J_0 = J_1 \cap J_2$. Our aim is to define a Fredholm module $x \diamond \widetilde{x}$ obtained, informally speaking, by "*pasting together along J_0 the part of x corresponding to J_1 with the part of $\widetilde{x}$ corresponding to J_2.*" The idea of the process is best illustrated by Fig. 2.4. More formally, let us represent $\widetilde{F}$ by the 3×3 block matrix

$$\widetilde{F} \approx \begin{pmatrix} \widetilde{a} & \widetilde{b} & 0 \\ \widetilde{b}^* & \widetilde{c} & \widetilde{d} \\ 0 & \widetilde{d}^* & \widetilde{e} \end{pmatrix}, \quad \widetilde{a} - \widetilde{a}^*, \quad \widetilde{c} - \widetilde{c}^*, \quad \widetilde{e} - \widetilde{e}^*, \qquad (2.13)$$

---

[1] Where 1 stands for the identity operators in the corresponding subspaces.

2.3. Superposition Principle

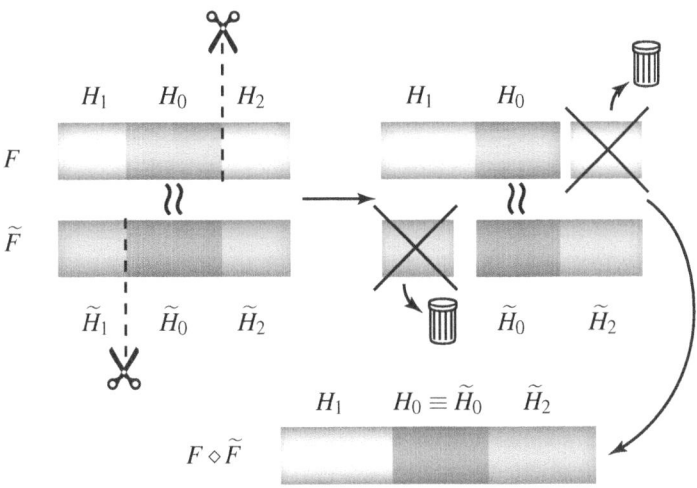

Figure 2.4: Cutting and pasting of Fredholm modules

by analogy with (2.10). The condition that x and $\widetilde{x}$ agree on J_0 acquires the form $Tc \approx \widetilde{c}T$. To simplify the notation, we identify H_0 with $\widetilde{H}_0$ with the use of T; then we can omit T altogether, and the condition of agreement on J_0 becomes $c \approx \widetilde{c}$. Set

$$H \diamond \widetilde{H} = H_1 \oplus H_0 \oplus \widetilde{H}_2, \quad \rho \diamond \widetilde{\rho} = \rho \mid H_1 \oplus H_0 \oplus \widetilde{\rho} \mid \widetilde{H}_2,$$

$$F \diamond \widetilde{F} = \begin{pmatrix} a & b & 0 \\ b^* & c & \widetilde{d} \\ 0 & \widetilde{d}^* & \widetilde{e} \end{pmatrix}. \qquad (2.14)$$

Proposition 2.6. *Formulas* (2.14) *specify a well-defined Fredholm module*

$$x \diamond \widetilde{x} = (\rho \diamond \widetilde{\rho}, H \diamond \widetilde{H}, F \diamond \widetilde{F})$$

over A.

Proof. It suffices to verify that $(F \diamond \widetilde{F})^2 \sim 1$. (The other conditions in (2.7) are obviously satisfied for $x \diamond \widetilde{x}$.) After squaring the matrix, the verification amounts to routine computations with the use of the relation $c \approx \widetilde{c}$ and Eqs. (2.11)–(2.12) for F and $\widetilde{F}$. For example, for the entry in the second column and second row we obtain

$$\varphi((F \diamond \widetilde{F})^2)_{22} = \varphi(b^*b + c^2 + \widetilde{d}\widetilde{d}^*) \sim \varphi_1(1-c^2) + \varphi c^2 + \varphi_2(1-c^2) = \varphi 1, \quad \varphi \in A. \quad \square$$

The Fredholm module $\widetilde{x} \diamond x$ is defined in a similar way.

Main result. Now we are in a position to state the main result of this chapter.

Theorem 2.7 (Superposition principle in K-homology). *Let A be a C^*-algebra, and let $J_1, J_2 \subset A$ be two ideals such that $J_1 + J_2 = A$. Next, let $x = (H, \rho, F)$ and $\tilde{x} = (\tilde{H}, \tilde{\rho}, \tilde{F})$ be nondegenerate Fredholm modules over A (graded or ungraded simultaneously) that agree on the ideal $J_0 = J_1 \cap J_2$ in the sense of Definition 2.5. Then*

$$[x \diamond \tilde{x}] - [x] = [\tilde{x}] - [\tilde{x} \diamond x], \tag{2.15}$$

where $[y] \in K^*(A)$ is the equivalence class of a Fredholm module y in the K-homology of A.

Relation (2.15) means that the difference of K-homology classes resulting from the disagreement of Fredholm modules on the ideal J_2 is independent of how the modules behave on J_1 (where they agree with each other).

Proof. It suffices to deform the Fredholm module $x \oplus \tilde{x}$ to a module that is unitarily equivalent to the module $(x \diamond \tilde{x}) \oplus (\tilde{x} \diamond x)$. The homotopy has the following structure. It is the family of Fredholm modules $(\rho \oplus \tilde{\rho}, H \oplus \tilde{H}, \mathscr{F}_t)$, $t \in [0, \pi/2]$, where the operator $\mathscr{F}_t$ is given in the decomposition

$$H \oplus \tilde{H} = H_1 \oplus H_0 \oplus H_2 \oplus \tilde{H}_1 \oplus H_0 \oplus \tilde{H}_2$$

by the 6×6 block matrix

$$\mathscr{F}_t = \begin{pmatrix} a & b & 0 & 0 & 0 & 0 \\ b^* & c & d\cos t & 0 & 0 & -\tilde{d}\sin t \\ 0 & d^*\cos t & e & 0 & d^*\sin t & 0 \\ 0 & 0 & 0 & \tilde{a} & \tilde{b} & 0 \\ 0 & 0 & d\sin t & \tilde{b}^* & c & \tilde{d}\cos t \\ 0 & -d^*\sin t & 0 & 0 & \tilde{d}^*\cos t & \tilde{e} \end{pmatrix}. \tag{2.16}$$

The first and second conditions in (2.7) are obviously satisfied for $\mathscr{F}_t$. To verify the third condition ($\mathscr{F}_t^2 \approx 1$), it suffices to analyze the 21 entries of the matrix $K = \mathscr{F}_t^2$ lying on or above the main diagonal with the use of the relation $\tilde{c} \approx c$, Eqs. (2.11)–(2.12), and their counterparts for $\tilde{F}$. Let us carry out these computations with necessary explanations:

$$K_{11} = a^2 + bb^* \approx 1, \quad K_{12} = ab + bc \approx 0, \quad K_{13} = bd\cos t \approx 0,$$
$$K_{14} = K_{15} = 0,$$
$$K_{16} = -b\tilde{d}\sin t \approx 0,$$

because $b\tilde{d} \approx 0$ by analogy with bd,

$$K_{22} = b^*b + c^2 + dd^*\cos^2 t + \widetilde{dd^*}\sin^2 t \approx 1,$$

because $\varphi K_{22} \sim \varphi_1(1-c^2) + c^2 + \varphi_2(1-c^2)(\cos^2 t + \sin^2 t) = \varphi 1$,

$$K_{23} = (cd + de)\cos t \approx 0, \quad K_{24} = 0,$$
$$K_{25} = (dd^* - \widetilde{dd^*})\cos t \sin t \approx 0,$$

2.3. Superposition Principle

because $\varphi K_{25} \sim (\varphi_2(1-c^2) - \varphi_2(1-c^2))\cos t \sin t = 0$,

$$K_{26} = -(c\widetilde{d} + \widetilde{de})\sin t \approx -(\widetilde{cd} + \widetilde{de})\sin t \approx 0$$
$$K_{33} = d^*d\cos^2 t + e^2 + d^*d\sin^2 t = d^*d + e^2 \approx 1,$$
$$K_{34} = d^*\widetilde{b}^* \sin t \approx 0$$

by analogy with K_{16},

$$K_{35} = (ed^* + d^*c)\sin t \approx 0,$$
$$K_{36} = (-d^*\widetilde{d} + d^*\widetilde{d})\sin t \cos t = 0,$$
$$K_{44} = \widetilde{a} + \widetilde{bb}^* \approx 1, \quad K_{45} = \widetilde{ab} + \widetilde{bc} \approx 0, \quad K_{46} = \widetilde{bd}\cos t \approx 0,$$
$$K_{55} = dd^*\sin^2 t + \widetilde{b}^*\widetilde{b} + c^2 + \widetilde{dd}^*\cos^2 t \approx 1$$

by analogy with K_{22},

$$K_{56} = (c\widetilde{d} + \widetilde{de})\cos t \approx (\widetilde{cd} + \widetilde{de})\cos t \approx 0,$$
$$K_{66} = \widetilde{d}^*\widetilde{d}(\sin^2 t + \cos^2 t) + \widetilde{e}^2 \approx 1.$$

We have proved that $\mathscr{F}_t^2 \approx 1$. Next, $\mathscr{F}_0 = F \oplus \widetilde{F}$ and

$$\mathscr{F}_{\pi/2} = \begin{pmatrix} a & b & 0 & 0 & 0 & 0 \\ b^* & c & 0 & 0 & 0 & -\widetilde{d} \\ 0 & 0 & e & 0 & d^* & 0 \\ 0 & 0 & 0 & \widetilde{a} & \widetilde{b} & 0 \\ 0 & 0 & d & \widetilde{b}^* & c & 0 \\ 0 & -\widetilde{d}^* & 0 & 0 & 0 & \widetilde{e} \end{pmatrix} = U^* \begin{pmatrix} a & b & 0 & 0 & 0 & 0 \\ b^* & c & \widetilde{d} & 0 & 0 & 0 \\ 0 & \widetilde{d}^* & \widetilde{e} & 0 & 0 & 0 \\ 0 & 0 & 0 & \widetilde{a} & \widetilde{b} & 0 \\ 0 & 0 & 0 & \widetilde{b}^* & c & d \\ 0 & 0 & 0 & 0 & d^* & e \end{pmatrix} U, \quad (2.17)$$

where the unitary operator

$$U: H \oplus \widetilde{H} \equiv H_1 \oplus H_0 \oplus H_2 \oplus \widetilde{H}_1 \oplus H_0 \oplus \widetilde{H}_2$$
$$\longrightarrow H_1 \oplus H_0 \oplus \widetilde{H}_2 \oplus \widetilde{H}_1 \oplus H_0 \oplus H_2 \equiv (H \diamond \widetilde{H}) \oplus (\widetilde{H} \diamond H)$$

interchanges the third and sixth components and multiplies the third component by -1. Thus, $\mathscr{F}_{\pi/2} = U^*((F \diamond \widetilde{F}) \oplus (\widetilde{F} \diamond F))U$, which completes the proof of the theorem. $\square$

Restatement of the main theorem. In applications, one often starts from the following situation: there is an "agreement diagram"

(2.18)

(see Definition 2.5) of Fredholm modules, and the underlying diagram

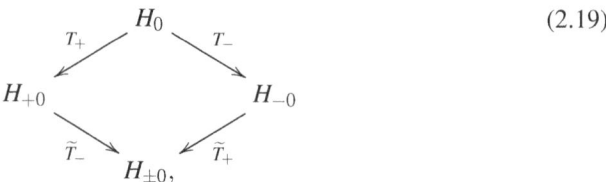
(2.19)

of the respective mappings restricted to the subspaces corresponding to the ideal J_0 commutes.

Theorem 2.8. *Under these conditions, one has*

$$[F_+] - [F] = [F_\pm] - [F_-].$$

Proof. One can readily see that if the diagram (2.19) commutes, then

$$F_+ \equiv F \diamond F_\pm, \qquad F_- \equiv F_\pm \diamond F.$$

It remains to use Theorem 2.7. □

2.4 Fredholm Modules and Elliptic Operators

Fredholm modules play the main role in all constructions related to K-homology of operator algebras including the superposition theorem, while it is *elliptic operators* that are of interest in applications. Thus, to apply the superposition theorem, one needs a procedure that would assign a Fredholm module to a given elliptic operator. The details of such a procedure, which is often referred to as *normalization*, depend on the problem considered and often constitute a crucial step in the solution. It is most suitable to discuss them when considering specific applications; here we give only a glimpse of how such a procedure can be done.

1. Let A be a unital C^*-algebra, and let $\rho : A \longrightarrow \mathfrak{B}(H)$ be a (unital) representation of A on a Hilbert space H. Next, let $D \in \mathfrak{B}(H)$ be a general elliptic operator in the sense of Atiyah [4], i.e., a Fredholm operator satisfying the locality property

$$[D, \rho(\varphi)] \sim 0 \qquad \forall \varphi \in A.$$

(The paper [4] deals with the case in which $A = C(X)$ for a compact space X.) Assume that D is self-adjoint, $D = D^*$. Then one can assign an ungraded Fredholm module (H, ρ, F) to D as follows. Since D is self-adjoint and Fredholm, it follows that zero is an isolated point of finite multiplicity of the spectrum of D, and hence we can write

$$1 = P_-(D) + P_0(D) + P_+(D),$$

2.4. Fredholm Modules and Elliptic Operators

where the $P_\pm(D)$ are the orthogonal projections in H onto the subspaces corresponding to the positive and negative parts $\sigma_\pm(D)$ of the spectrum of D and P_0 is the (finite rank) projection onto the null space of D. Now set

$$F = \operatorname{sign} D \stackrel{\text{def}}{=} P_+(D) - P_-(D).$$

Let us verify that (H, ρ, F) is a Fredholm module. One obviously has $F = F^*$, and

$$F^2 = P_+(D) + P_-(D) = 1 - P_0(D) \sim 1,$$

because $P_0(D)$ is finite rank. To prove that F compactly commutes with the action of A, consider an odd function $f \in C_0^\infty(\mathbb{R})$ such that

$$f(\tau) = \begin{cases} 1, & \tau \in \sigma_+(D), \\ -1, & \tau \in \sigma_-(D). \end{cases}$$

The existence of such a function follows from the fact that some deleted neighborhood $(-\varepsilon, \varepsilon) \setminus \{0\}$ of zero does not contain points of spectrum of D; see Fig. 2.5. We have

$$F = f(D) = \frac{1}{\sqrt{2\pi}} \int_{-\infty}^\infty \tilde{f}(t) e^{iDt} dt,$$

where $\tilde{f}$ is the Fourier transform of f and the integral converges absolutely in the operator norm. Let $B = \rho(\varphi)$, $\varphi \in A$. Using the identity

$$e^{iDt} B - B e^{iDt} = \int_0^1 \frac{d}{d\theta} e^{iD\theta t} B e^{iD(1-\theta)t} d\theta = it \int_0^1 \frac{d}{d\theta} e^{iD\theta t} [D, B] e^{iD(1-\theta)t} d\theta,$$

we represent the commutator $[F, B]$ as

$$[F, B] = \frac{i}{\sqrt{2\pi}} \int_{-\infty}^\infty \int_0^1 t \tilde{f}(t) e^{iD\theta t} [D, B] e^{iD(1-\theta)t} d\theta \, dt.$$

The integrand is a compact operator for any (θ, t), and the integral converges absolutely in the operator norm, because $\tilde{f}$ is smooth and rapidly decays at infinity. Thus, $[F, \rho(\varphi)] \sim 0$, as desired, and (H, ρ, F) is indeed an ungraded Fredholm module.

2. Now let $\rho_1 : A \longrightarrow \mathfrak{B}(H_1)$ and $\rho_2 : A \longrightarrow \mathfrak{B}(H_2)$ be two representations of A on Hilbert spaces H_1 and H_2, respectively, and let $C \in \mathfrak{B}(H_1, H_2)$ be a general elliptic operator in the sense of Atiyah [4]; the locality property reads

$$\rho_2(\varphi) C - C \rho_1(\varphi) \sim 0 \qquad \forall \varphi \in A.$$

To the operator C, we assign a graded Fredholm module (H, ρ, F) as follows. Set $H = H_1 \oplus H_2$ and interpret H as a $\mathbb{Z}_2$-graded space with $H_+ = H_1$ and $H_- = H_2$. Next, set $\rho = \rho_1 \oplus \rho_2$; that is, in the block 2×2 matrix form the representation ρ is given by

$$\rho(\varphi) = \begin{pmatrix} \rho_1(\varphi) & 0 \\ 0 & \rho_2(\varphi) \end{pmatrix}, \qquad \varphi \in A.$$

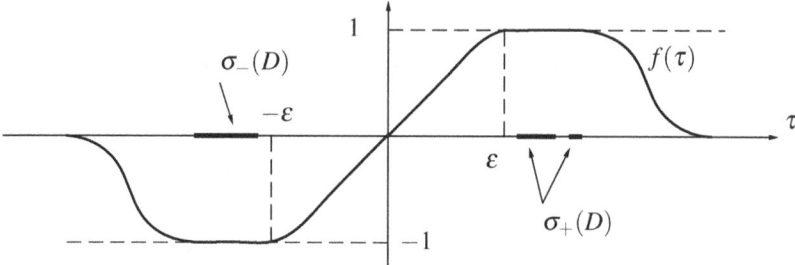

Figure 2.5: The function $f(\tau)$

The operator

$$D = \begin{pmatrix} 0 & C^* \\ C & 0 \end{pmatrix}$$

is general elliptic and self-adjoint, and we set, by the same formula as in the ungraded case,

$$F = P_+(D) - P_-(D).$$

The properties $F^* = F$ and $F^2 - 1 \sim 0$ are obvious by the preceding, and we only need to verify that F is odd with respect to the grading. This is equivalent to saying that F anti-commutes with the involution

$$\alpha = \begin{pmatrix} 1 & 0 \\ 0 & -1 \end{pmatrix}.$$

To prove this, note that $F = f(D)$, where $f(\tau)$ is a smooth odd function (see Fig 2.5) and hence

$$f(\tau) = \tau g(\tau^2),$$

where $g(z)$ is again a smooth function. Accordingly, we have

$$f(D) = Dg(D^2) = \begin{pmatrix} 0 & C^* \\ C & 0 \end{pmatrix} \begin{pmatrix} g(C^*C) & 0 \\ 0 & g(CC^*) \end{pmatrix}$$

and

$$\begin{aligned}\alpha f(D) &= \begin{pmatrix} 1 & 0 \\ 0 & -1 \end{pmatrix} \begin{pmatrix} 0 & C^* \\ C & 0 \end{pmatrix} \begin{pmatrix} g(C^*C) & 0 \\ 0 & g(CC^*) \end{pmatrix} \\ &= -\begin{pmatrix} 0 & C^* \\ C & 0 \end{pmatrix} \begin{pmatrix} 1 & 0 \\ 0 & -1 \end{pmatrix} \begin{pmatrix} g(C^*C) & 0 \\ 0 & g(CC^*) \end{pmatrix} \\ &= -\begin{pmatrix} 0 & C^* \\ C & 0 \end{pmatrix} \begin{pmatrix} g(C^*C) & 0 \\ 0 & g(CC^*) \end{pmatrix} \begin{pmatrix} 1 & 0 \\ 0 & -1 \end{pmatrix} = -f(D)\alpha,\end{aligned}$$

as desired.

2.4. Fredholm Modules and Elliptic Operators

3. It is instructive to write out the operator F in a more explicit form. We have
$$\tau g(\tau^2) = \operatorname{sign} \tau \quad \text{on} \quad \sigma_+(D) \cup \sigma_-(D)$$
and hence
$$g(z) = z^{-1/2} \quad \text{on} \quad \sigma(D^2) \setminus \{0\}.$$
Thus,
$$F = D(D^2)^{[-1/2]} = (D^2)^{[-1/2]}D = D|D|^{[-1]} = |D|^{[-1]}D,$$
where $(D^2)^{[-1/2]} = |D|^{[-1]}$ is a regularized power of D^2,
$$(D^2)^{[-1/2]} = \begin{cases} 0 & \text{on } \ker D, \\ (D^2|_{(\ker D)^\perp})^{-1/2} & \text{on } (\ker D)^\perp. \end{cases}$$

(This is well defined, because the spectrum of $D^2|_{(\ker D)^\perp}$ is bounded away from zero.)
For the graded case, we can further write
$$F = \begin{pmatrix} 0 & C^*(CC^*)^{[-1/2]} \\ C(C^*C)^{[-1/2]} & 0 \end{pmatrix} = \begin{pmatrix} 0 & (C^*C)^{[-1/2]}C^* \\ (CC^*)^{[-1/2]}C & 0 \end{pmatrix}, \quad (2.20)$$
where
$$(C^*C)^{[-1/2]} = \begin{cases} 0 & \text{on } \ker C, \\ (C^*C|_{(\ker C)^\perp})^{-1/2} & \text{on } (\ker C^*C)^\perp \end{cases}$$
and a similar definition is used for $(CC^*)^{[-1/2]}$.

4. Elliptic operators occurring in applications usually do not directly fall within the simple scheme described above. For example, if we consider an elliptic differential operator C of order s on a smooth closed manifold M, then this operator can of course be treated as a bounded operator, say, in the spaces $C \colon H^s(M, E_1) \longrightarrow L_2(M, E_2)$ of sections of vector bundles on M, but the Sobolev space $H^s(M, E_1)$ does not bear an action of the C^*-algebra $C(M)$, and the normalization should involve order reduction, so that we end up with operators acting in L_2 spaces. Thus, in applications one deals with variations on the theme (2.20). On a smooth closed manifold, one can take the formula
$$F = \begin{pmatrix} 0 & C^*(CC^*)^{[-1/2]} \\ C(C^*C)^{[-1/2]} & 0 \end{pmatrix}, \quad (2.21)$$
which formally coincides with (2.20) but treats C as an *unbounded operator* acting from $L_2(M, E_1)$ to $L_2(M, E_2)$, or the slightly more convenient formula
$$F = \begin{pmatrix} 0 & C^*(1+CC^*)^{-1/2} \\ C(1+C^*C)^{-1/2} & 0 \end{pmatrix}. \quad (2.22)$$

Note that formula (2.22) *does not give the right answer in general*. The difference between (2.22) and (2.21) is a compact operator because it so happens that, for the case of a smooth closed manifold, the operator C^*C (as well as CC^*) is an operator with compact resolvent (or, putting this differently, the embedding of $H^{2s}(M,E_j)$ in $L_2(M,E_j)$ is a compact operator). This is not the case, for example, on noncompact manifolds, and so different tools should be used there.

Chapter 3

Superposition Principle for KK-Theory

3.1 Preliminaries

In the present chapter, we prove a relative index type theorem that compares certain elements of the Kasparov KK-group $KK(A,B)$. Just as the superposition principle in Chapter 2 permits one to obtain relative index theorems for elliptic operators, the theorem given here implies relative index theorems for elliptic operators over a C^*-algebra B in the sense of Mishchenko–Fomenko [47], where the index is an element of the K-group $K_0(B)$. Although it might seem at first glance that the results in Chapter 2 should be a special case of those given here for $B = \mathbb{C}$, this is actually not the case. More precisely, the superposition theorem of Chapter 2 holds for unital as well as nonunital algebras A, whereas here we are only able to consider the case of unital algebras A. This is because submodules of Hilbert modules, unlike subspaces of Hilbert spaces, are not always complemented, and so the constructions of Chapter 2 cannot be used; one needs a completely different technique, and so far we have one only for the unital case.

The results of this chapter were obtained in [62]. The material is a little more advanced than in the rest of the book. We freely use notions and notation related to C^*-algebras and KK-theory (e.g., see [13, 35, 66] and the literature cited therein), although we do recall a limited amount of basic information in Section 3.2.

3.2 Hilbert Modules, Kasparov Modules, and KK

Hilbert modules. Let B be a C^*-algebra.

Definition 3.1. A *pre-Hilbert module* over B is a right B-module H equipped with a

B-valued inner product, i.e., a sesquilinear[1] mapping

$$\langle \cdot, \cdot \rangle_H : H \times H \longrightarrow B,$$
$$(\xi, \eta) \longmapsto \langle \xi, \eta \rangle_H$$

with the following properties:

(i) $\langle \xi, \eta b \rangle_H = \langle \xi, \eta \rangle_H b, \quad b \in B.$
(ii) $\langle \xi, \eta \rangle_H = \langle \eta, \xi \rangle_H^*.$
(iii) $\langle \xi, \xi \rangle_H \geq 0; \quad \langle \xi, \xi \rangle_H = 0 \implies \xi = 0.$

The function

$$\|\xi\|_H = \|\langle \xi, \xi \rangle_H\|_B^{1/2}$$

satisfies the axioms of norm; if H is complete in this norm (i.e., is a Banach space), then H is called a *Hilbert module* over B, or a Hilbert B-module.

Let H be a Hilbert B-module, and let $C : H \longrightarrow H$ be a bounded linear operator such that

$$C(\xi b) = (C\xi)b, \quad \xi \in H, \quad b \in B \quad (B\text{-linearity}).$$

If there exists a bounded B-linear operator $C^* : H \longrightarrow H$ such that

$$\langle \xi, C\eta \rangle_H = \langle C^*\xi, \eta \rangle_H, \quad \xi, \eta \in H,$$

then C is said to be *adjointable*. Bounded B-linear adjointable operators form a unital C^*-algebra, which will denoted by $\mathbb{B}(H)$. Operators C_{η_1, η_2} of the form

$$C_{\eta_1, \eta_2} \xi = \eta_2 \langle \eta_1, \xi \rangle_H$$

belong to $\mathbb{B}(H)$; they are called *rank 1 operators*. The closed linear span of rank 1 operators is a closed ideal in $\mathbb{B}(H)$; this ideal will be denoted by $\mathbb{K}(H)$ and is called the ideal of B-compact operators. For $C, D \in \mathbb{B}(H)$, we write $C \sim D$ if $C - D \in \mathbb{K}(H)$.

If H_1 and H_2 are Hilbert B-modules, then one defines the space $\mathbb{B}(H_1, H_2)$ of adjointable operators and the subspace $\mathbb{K}(H_1, H_2)$ of B-compact operators in a completely similar way.

Kasparov modules. Let A and B be graded[2] C^*-algebras.

Definition 3.2. A *Kasparov module* for (A, B) is a triple (H, ρ, F) consisting of a graded countably generated Hilbert module H over B, a graded $*$-homomorphism $\rho : A \to \mathbb{B}(H)$ of A into the C^*-algebra $\mathbb{B}(H)$ of adjointable operators on H, and an operator $F \in \mathbb{B}(H)$ of degree 1 such that

$$[F, \rho(a)] \sim 0, \quad \rho(a)(F^2 - 1) \sim 0, \quad \rho(a)(F - F^*) \sim 0 \quad (3.1)$$

for every $a \in A$.

A *degenerate* Kasparov module is one in which all relations in (3.1) are satisfied exactly rather than modulo B-compact operators.

---

[1] Antilinear in the first argument and linear in the second.
[2] By graded we always mean $\mathbb{Z}_2$-graded. The case in which $A_1 = 0$ and/or $B_1 = 0$ is not excluded.

3.3. Superposition Principle

KK-theory. The elements of Kasparov KK-groups $KK(A,B)$ are defined as equivalence classes of Kasparov (A,B)-modules for an elaborate equivalence relation. We do not go into much detail here, referring the reader, e.g., to [13, Chapter VIII], and only mention two cases, actually needed in our proofs, in which two Kasparov (A,B)-modules $x_0 = (H_0, \rho_0, F_0)$ and $x_1 = (H_1, \rho_1, F_1)$ are equivalent.

1. The modules x_0 and x_1 are equivalent if there exists a unitary $U \in \mathbb{B}(H_0, H_1)$ of degree 0 that intertwines the ρ_j and F_j,

$$U\rho_0 = \rho_1 U, \qquad UF_0 = F_1 U.$$

This is just the *unitary equivalence* $\approx_u$ in [13, Definition 17.2.1].

2. The modules x_0 and x_1 are equivalent if $H_0 = H_1$, $F_0 = F_1$, and the homomorphisms ρ_0 and ρ_1 can be included in a family ρ_t, $t \in [0,1]$, of graded $*$-homomorphisms such that each (H_0, ρ_t, F_0) is a Kasparov module and $\rho_t(a)$ is norm continuous in t for each $a \in A$. This is a special case of a *standard homotopy* in [13, Definition 17.2.1].

Remark 3.3. In what follows, to make the text more readable and avoid unnecessary technicalities, we generally say nothing about the grading, tacitly assuming that all elements, operators, homomorphisms, etc. involved in our argument have appropriate degrees. The verification of this fact, which might be boring, is left to the reader.

3.3 Superposition Principle

Algebra A and a partition of unity. We will work in the following setting. Let A be a unital C^*-algebra, and let $J_1, J_2 \subset A$ be two ideals such that

$$J_1 + J_2 = A.$$

By J_0 we denote the intersection of these ideals, $J_0 = J_1 \cap J_2$. We need the following technical lemma.

Lemma 3.4. *There exist self-adjoint positive elements $\psi_1 \in J_1$ and $\psi_2 \in J_2$ with*

$$\psi_1^2 + \psi_2^2 = 1, \qquad [\psi_1, \psi_2] = 0. \tag{3.2}$$

Proof. Since $J_1 + J_2 = A$, we see that there exists an element $\chi \in J_1$ such that $1 - \chi \in J_2$; taking the real part, we can assume that $\chi = \chi^*$. By functional calculus, the element $\chi^2 + (1-\chi)^2$ is positive and invertible, and we set

$$\psi_1 = |\chi|(\chi^2 + (1-\chi)^2)^{-1/2}, \qquad \psi_2 = |1-\chi|(\chi^2 + (1-\chi)^2)^{-1/2}. \qquad \square$$

Structure of Kasparov (A,B)-modules. We only consider Kasparov modules in which the homomorphism ρ is unital (and refer to these as *unital* Kasparov modules); in this case, the factor $\rho(a)$ can be dropped in the second and third conditions in (3.1). Let $x = (H, \rho, F)$ be such a module. The (A,B)-sub-bimodules[3]

$$H_1 = \rho[J_1]H, \qquad H_2 = \rho[J_2]H, \qquad H_0 = \rho[J_0]H \tag{3.3}$$

---
[3] Note that our notation differs from that used in Chapter 2.

of H are closed (the proof mimics that in [35, Exercise 1.9.17]) and hence are simultaneously Hilbert B-modules.

Lemma 3.5. *One has $H_0 = H_1 \cap H_2$.*

Proof. The inclusion $H_0 \subset H_1 \cap H_2$ is obvious. To prove the converse inclusion, let $\xi \in H_1 \cap H_2$. Let $\{u_\lambda\}$ and $\{v_\mu\}$ be approximate units [66, 1.4.1] for J_1 and J_2, respectively. Then $\rho(u_\lambda)\xi \to \xi$ and $\rho(v_\mu)\xi \to \xi$. We have

$$\rho(u_\lambda v_\mu)\xi - \xi = \rho(u_\lambda)(\rho(v_\mu)\xi - \xi) + \rho(u_\lambda)\xi - \xi$$

and hence

$$\|\rho(u_\lambda v_\mu)\xi - \xi\| \leq \|\rho(u_\lambda)\| \|\rho(v_\mu)\xi - \xi\| + \|\rho(u_\lambda)\xi - \xi\|$$
$$\leq \|\rho(v_\mu)\xi - \xi\| + \|\rho(u_\lambda)\xi - \xi\|.$$

It follows that there exists a subsequence of $\{\rho(u_\lambda v_\mu)\xi\}$ that converges to ξ, and hence $\xi \in H_0$, because $u_\lambda v_\mu \in J_0$ and $\rho(u_\lambda v_\mu)\xi \in H_0$. $\square$

Lemma 3.6. *The (A,B)-bimodule H is naturally isomorphic to the quotient $(H_1 \oplus H_2)/\Delta$, where*

$$\Delta = \{(\xi_1, \xi_2) \in H_1 \oplus H_2 : \xi_1 = -\xi_2 \in H_0\}.$$

Proof. Consider the (A,B)-bimodule homomorphism

$$\alpha \colon H_1 \oplus H_2 \longrightarrow H, \qquad (\xi_1, \xi_2) \longmapsto \xi_1 + \xi_2.$$

It is easily seen that $\alpha|_\Delta = 0$, and hence α descends to an (A,B)-bimodule homomorphism

$$\widehat{\alpha} \colon (H_1 \oplus H_2)/\Delta \longrightarrow H.$$

Now consider the mapping

$$\widehat{\beta} = \pi \circ \beta \colon H \longrightarrow (H_1 \oplus H_2)/\Delta,$$

where

$$\pi \colon H_1 \oplus H_2 \longrightarrow (H_1 \oplus H_2)/\Delta$$

is the natural projection and β is given by

$$\beta \colon H \longrightarrow H_1 \oplus H_2, \qquad \xi \longmapsto (\rho(\psi_1^2)\xi, \rho(\psi_2^2)\xi).$$

We have

$$\widehat{\alpha} \circ \widehat{\beta}(\xi) = \alpha \circ \beta(\xi) = \rho(\psi_1^2)\xi + \rho(\psi_2^2)\xi = \xi$$

and

$$\widehat{\beta} \circ \widehat{\alpha}(\pi(\xi_1, \xi_2)) = \widehat{\beta}(\xi_1 + \xi_2) = \pi(\rho(\psi_1^2)(\xi_1 + \xi_2), \rho(\psi_2^2)(\xi_1 + \xi_2))$$
$$= \pi(\xi_1, \xi_2) + \pi(\rho(\psi_1^2)\xi_2 - \rho(\psi_2^2)\xi_1, \rho(\psi_2^2)\xi_1 - \rho(\psi_1^2)\xi_2) = \pi(\xi_1, \xi_2),$$

because $(\rho(\psi_1^2)\xi_2 - \rho(\psi_2^2)\xi_1, \rho(\psi_2^2)\xi_1 - \rho(\psi_1^2)\xi_2) \in \Delta$. Thus, $\widehat{\beta}$ is the two-sided inverse of $\widehat{\alpha}$, hence an (A,B)-bimodule homomorphism, and the proof is complete. $\square$

3.3. Superposition Principle

Cutting and pasting. Let $\tilde{x} = (\tilde{H}, \tilde{\rho}, \tilde{F})$ be another unital Kasparov (A,B)-module, and let $\tilde{H}_j$, $j = 0,1,2$, be the (A,B)-sub-bimodules of $\tilde{H}$ defined as in (3.3).

Definition 3.7. We say that x and $\tilde{x}$ *agree on* J_0 if there is a unitary (in the sense of Hilbert modules over B) isomorphism $T \colon H_0 \to \tilde{H}_0$ of (A,B)-bimodules such that, for arbitrary $c, d \in J_0$, one has
$$T\rho(c)F\rho(d) \sim \tilde{\rho}(c)\tilde{F}\tilde{\rho}(d)T. \tag{3.4}$$
(Note that $\rho(c)F\rho(d) \in \mathbb{B}(H_0)$ and $\tilde{\rho}(c)\tilde{F}\tilde{\rho}(d) \in \mathbb{B}(\tilde{H}_0)$ are well defined.)

Assume that x and $\tilde{x}$ agree on J_0. Our aim is to use some sort of *cutting-and-pasting procedure* to define a unital Kasparov (A,B)-module $x \diamond \tilde{x}$ that agrees with x on J_1 and with $\tilde{x}$ on J_2. To this end, consider the (A,B)-bimodule
$$H \diamond \tilde{H} = (H_1 \oplus \tilde{H}_2) \big/ \{(\xi_1, \xi_2) \colon \xi_1 \in H_0,\ \xi_2 \in \tilde{H}_0,\ T\xi_1 + \xi_2 = 0\}. \tag{3.5}$$
The elements of $H \diamond \tilde{H}$ will be denoted by $\xi = [(\xi_1, \xi_2)]$, and the action of A on $H \diamond \tilde{H}$ will be denoted by $\rho \diamond \tilde{\rho}$,
$$(\rho \diamond \tilde{\rho})(\varphi)\xi = [(\rho(\varphi)\xi_1, \tilde{\rho}(\varphi)\xi_2)].$$
Note that H_1 and $\tilde{H}_2$ are naturally embedded in $H \diamond \tilde{H}$ (the embeddings are induced by those of H_1 and $\tilde{H}_2$ in the direct sum $H_1 \oplus \tilde{H}_2$), and if we identify H_1 and $\tilde{H}_2$ with their images under these embeddings,
$$H_1 \simeq (H \diamond \tilde{H})_1 \equiv (\rho \diamond \tilde{\rho})[J_1](H \diamond \tilde{H}), \quad \tilde{H}_2 \simeq (H \diamond \tilde{H})_2 \equiv (\rho \diamond \tilde{\rho})[J_2](H \diamond \tilde{H}), \tag{3.6}$$
then, for arbitrary $\xi \in H \diamond \tilde{H}$ and $\varphi_j \in J_j$, $j = 1, 2$, we have
$$(\rho \diamond \tilde{\rho})(\varphi_1)\xi \simeq \rho(\varphi_1)\xi_1 + T^*\tilde{\rho}(\varphi_1)\xi_2 \in H_1,$$
$$(\rho \diamond \tilde{\rho})(\varphi_2)\xi \simeq T\rho(\varphi_2)\xi_1 + \tilde{\rho}(\varphi_2)\xi_2 \in \tilde{H}_2.$$

From now on, to simplify the notation, we identify H_0 and $\tilde{H}_0$ via T, accordingly suppress T in all the formulas, and also write simply φ instead of $\rho(\varphi)$, $\tilde{\rho}(\varphi)$, or $(\rho \diamond \tilde{\rho})(\varphi)$. This will not lead to a misunderstanding even if several representations are involved, because which is meant is always clear from the context.

Lemma 3.8. *The formula*
$$\langle \xi, \eta \rangle_{H \diamond \tilde{H}} = \langle \psi_1^2 \xi, \psi_1^2 \eta \rangle_H + \langle \psi_2^2 \xi, \psi_2^2 \eta \rangle_{\tilde{H}} + 2\langle \psi_1 \psi_2 \xi, \psi_1 \psi_2 \eta \rangle_H, \tag{3.7}$$
where $\langle \cdot, \cdot \rangle_H$ and $\langle \cdot, \cdot \rangle_{\tilde{H}}$ are the B-valued inner products on H and $\tilde{H}$, respectively, specifies a well-defined B-valued inner product on $H \diamond \tilde{H}$, which makes $H \diamond \tilde{H}$ a Hilbert B-module and the action of A on $H \diamond \tilde{H}$ a unital $*$-homomorphism $\rho \diamond \tilde{\rho} \colon A \to \mathbb{B}(H \diamond \tilde{H})$.

Proof. Note that the right-hand side of (3.7) is well defined. Indeed, by our identifications, $\psi_1^2 \xi, \psi_1^2 \eta \in H_1 \subset H$, because $\psi_1^2 \in J_1$, and so the inner product of these elements in H makes sense. Likewise, the second summand is well defined. As for the last summand, we have $\psi_1\psi_2 \in J_0$ and hence $\psi_1\psi_2\xi, \psi_1\psi_2\eta \in H_0 \subset H$, and the inner product again makes sense. Properties (i)–(iii) in the definition of a B-valued inner product are obviously satisfied; for example, if $\langle \xi, \xi \rangle_{H \diamond \widetilde{H}} = 0$, then $\langle \psi_1^2 \xi, \psi_1^2 \xi \rangle_H = 0$ and $\langle \psi_2^2 \xi, \psi_2^2 \xi \rangle_{\widetilde{H}} = 0$. It follows that $\psi_1^2 \xi = 0$, $\psi_2^2 \xi = 0$, and hence $\xi = \psi_1^2 \xi + \psi_2^2 \xi = 0$.

Now let us prove that $\rho \diamond \widetilde{\rho}$ is indeed a $*$-homomorphism with respect to the inner product (3.7). For brevity, let us for now denote this inner product simply by $\langle \cdot, \cdot \rangle$. Thus, we need to prove that
$$\langle a\xi, \eta \rangle = \langle \xi, a^*\eta \rangle \qquad \forall a \in A.$$
Since the H- and $\widetilde{H}$-inner products coincide on $H_0 \equiv \widetilde{H}_0 = H_1 \cap \widetilde{H}_2$, it follows by routine computations that
$$\langle \xi, \eta \rangle = \begin{cases} \langle \xi, \eta \rangle_1 & \text{(the H-inner product restricted to H_1) if } \xi, \eta \in H_1, \\ \langle \xi, \eta \rangle_2 & \text{(the $\widetilde{H}$-inner product restricted to $\widetilde{H}_2$) if } \xi, \eta \in \widetilde{H}_2, \\ \langle \xi, \eta \rangle_0 & (= \langle \xi, \eta \rangle_1 = \langle \xi, \eta \rangle_2) \text{ if } \xi, \eta \in H_0. \end{cases}$$

Let $\{w_\lambda\}$ be an approximate unit for the ideal J_0. We claim that, for any $\xi, \eta \in H \diamond \widetilde{H}$ and $a \in A$, one has
$$\langle a\xi, \eta \rangle - \langle \xi, a^*\eta \rangle = \lim_\lambda \big(\langle aw_\lambda \xi, w_\lambda \eta \rangle - \langle w_\lambda \xi, a^* w_\lambda \eta \rangle \big). \tag{3.8}$$

Suppose momentarily that this is true. Then we are done. Indeed, $w_\lambda \xi, w_\lambda \eta \in H_0$ and hence
$$\langle aw_\lambda \xi, w_\lambda \eta \rangle - \langle w_\lambda \xi, a^* w_\lambda \eta \rangle = \langle aw_\lambda \xi, w_\lambda \eta \rangle_0 - \langle w_\lambda \xi, a^* w_\lambda \eta \rangle_0 = 0,$$
because, say, ρ is a $*$-homomorphism. Thus, it remains to prove (3.8). The difference on the left-hand side in (3.8) can be represented as the sum of three differences corresponding to the respective terms on the right-hand side in (3.7). Consider the first term. (The argument for the other two terms is similar.) We have
$$\langle \psi_1^2 a \xi, \psi_1^2 \eta \rangle_1 - \langle \psi_1^2 \xi, \psi_1^2 a^* \eta \rangle_1 = \langle [\psi_1^2, a] \xi, \psi_1^2 \eta \rangle_1 - \langle \psi_1^2 \xi, [\psi_1^2, a^*] \eta \rangle_1, \tag{3.9}$$
where the brackets stand for the commutator. Now
$$[\psi_1^2, a] = -[\psi_2^2, a],$$
and we conclude that $[\psi_1^2, a] \in J_1 \cap J_2 = J_0$. Hence
$$[\psi_1^2, a] = \lim_\lambda [\psi_1^2, a] w_\lambda,$$
and we obtain
$$\langle [\psi_1^2, a] \xi, \psi_1^2 \eta \rangle_1 = \lim_\lambda \langle [\psi_1^2, a] w_\lambda \xi, \psi_1^2 \eta \rangle_1 = -\lim_\lambda \langle w_\lambda \xi, [\psi_1^2, a^*] \psi_1^2 \eta \rangle_1.$$
Since $[\psi_1^2, a^*] \psi_1^2 \in J_0$, we have $[\psi_1^2, a^*] \psi_1^2 w_\lambda \to [\psi_1^2, a^*] \psi_1^2$. We conclude that
$$\langle [\psi_1^2, a] \xi, \psi_1^2 \eta \rangle_1 = -\lim_\lambda \langle w_\lambda \xi, [\psi_1^2, a^*] \psi_1^2 w_\lambda \eta \rangle_1 = \lim_\lambda \langle [\psi_1^2, a] w_\lambda \xi, \psi_1^2 w_\lambda \eta \rangle_1.$$

The subtrahend in (3.9) can be handled in a similar way, and the proof is complete. $\square$

3.3. Superposition Principle

Now set
$$F \diamond \widetilde{F} = \psi_1 F \psi_1 + \psi_2 \widetilde{F} \psi_2 : H \diamond \widetilde{H} \longrightarrow H \diamond \widetilde{H}. \tag{3.10}$$

This is well defined. Indeed, for example, if $\xi \in H \diamond \widetilde{H}$, then, in view of our identifications, $\psi_1 \xi \in (H \diamond \widetilde{H})_1 = H_1 \subset H$, hence $F \psi_1 \xi$ is a well-defined element of H, and hence $\psi_1 F \psi_1 \xi \in H_1 \subset H \diamond \widetilde{H}$ is a well-defined element of $H \diamond \widetilde{H}$.

Theorem 3.9. *The triple* $x \diamond \widetilde{x} = (H \diamond \widetilde{H}, \rho \diamond \widetilde{\rho}, F \diamond \widetilde{F})$ *is a unital Kasparov* (A, B)*-module, which is independent modulo B-compact perturbations of the choice of the partition of unity* (3.2) *and agrees with x on J_1 and with $\widetilde{x}$ on J_2.*

Proof. **1.** Let us prove that $(F \diamond \widetilde{F})^* \sim F \diamond \widetilde{F}$. We have
$$F \diamond \widetilde{F} - (F \diamond \widetilde{F})^* = \psi_1 (F - F^*) \psi_1 + \psi_2 (\widetilde{F} - \widetilde{F}^*) \psi_2.$$

The first term can be represented as the composition of operators
$$H \diamond \widetilde{H} \xrightarrow{\psi_1} H_1 \subset H \xrightarrow{F - F^*} H \xrightarrow{\psi_1} H_1 \subset H \diamond \widetilde{H}$$

and is B-compact, because so is $F - F^*$. A similar computation can be used for the second term.

2. Let us prove that $(F \diamond \widetilde{F})^2 \sim 1$. We have
$$(F \diamond \widetilde{F})^2 = \psi_1 F \psi_1^2 F \psi_1 + \psi_2 \widetilde{F} \psi_2^2 \widetilde{F} \psi_2 + \psi_1 F \psi_1 \psi_2 \widetilde{F} \psi_2 + \psi_2 \widetilde{F} \psi_2 \psi_1 F \psi_1. \tag{3.11}$$

Now (omitting details like in **1**)
$$\psi_1 F \psi_1^2 F \psi_1 \sim \psi_1 F^2 \psi_1^3 \psi_1 \sim \psi_1^4, \qquad \psi_2 \widetilde{F} \psi_2^2 \widetilde{F} \psi_2 \sim \psi_2^4.$$

The other two terms require a subtler approach. We cannot just commute, say, $\widetilde{F}$ with ψ_2 modulo B-compact operators, because expressions like $\psi_1 \widetilde{F} \psi_2$ make no sense in $H \diamond \widetilde{H}$. Instead, note that since ψ_1 and ψ_2 are positive and commute, we have
$$\psi_1 \psi_2 = d^2, \qquad d = \psi_1^{1/2} \psi_2^{1/2} \in J_0.$$

Consequently,
$$\psi_1 F \psi_1 \psi_2 \widetilde{F} \psi_2 = \psi_1 F d^2 \widetilde{F} \psi_2 \sim \psi_1 F d \widetilde{F} d \psi_2$$
$$\sim \psi_1 F d F d \psi_2 \sim \psi_1 F^2 d^2 \psi_2 \sim \psi_1 d^2 \psi_2 = \psi_1^2 \psi_2^2,$$

and a similar computation holds for the last term in (3.11). All in all, we obtain
$$(F \diamond \widetilde{F})^2 \sim \psi_1^4 + \psi_2^4 + 2 \psi_1^2 \psi_2^2 = 1.$$

3. Let us prove that $[F \diamond \widetilde{F}, a] \sim 0$ for any $a \in A$. We have
$$[F \diamond \widetilde{F}, a] = \psi_1 [F, a] \psi_1 + \psi_2 [\widetilde{F}, a] \psi_2$$
$$+ [\psi_1, a] F \psi_1 + [\psi_2, a] \widetilde{F} \psi_2 + \psi_1 F [\psi_1, a] + \psi_2 \widetilde{F} [\psi_2, a].$$

The first two terms are B-compact. Let $C = C_1 + C_2 + C_3 + C_4$ be the sum of the remaining terms. We claim that
$$C = \lim_\lambda (w_\lambda C w_\lambda + K_\lambda), \qquad (3.12)$$
where $\{w_\lambda\}$ is an approximate unit for J_0 and the K_λ are some B-compact operators. Indeed, let us show this, say, for C_1. We have $[a, \psi_1] \in J_1$ and
$$[a, \psi_1] = [a, (1-\psi_2^2)^{1/2}] = [a, (1-\psi_2^2)^{1/2} - 1] \in J_2$$
by functional calculus, because the function $(1-\psi_2^2)^{1/2} - 1$, which is well defined and continuous on the spectrum $\sigma(\psi_2) \subset [0,1]$, is zero at $\tau = 0$. We conclude that $[\psi_1, a] \in J_0 = J_1 \cap J_2$. Consequently,
$$[\psi_1, a] F \psi_1 = \lim_\lambda w_\lambda [\psi_1, a] F \psi_1 = \lim_\lambda \big(w_\lambda F [\psi_1, a] \psi_1 + w_\lambda [[\psi_1, a], F] \psi_1\big)$$
$$= \lim_\lambda \big(w_\lambda F [\psi_1, a] \psi_1 w_\lambda + w_\lambda [[\psi_1, a], F] \psi_1\big),$$
because $\|w_\lambda\| \leq 1$ and $[\psi_1, a] \psi_1 (1 - w_\lambda) \to 0$. Continuing the computation, we obtain
$$[\psi_1, a] F \psi_1 = \lim_\lambda \big(w_\lambda [\psi_1, a] F \psi_1 w_\lambda + w_\lambda [[\psi_1, a], F] \psi_1 (1 - w_\lambda)\big),$$
where the operator $w_\lambda [[\psi_1, a], F] \psi_1 (1-w_\lambda)$ is B-compact, because so is the double commutator. The computations for the other C_j are similar. It remains to show that $w_\lambda C w_\lambda$ is B-compact; then (3.12) implies that so is C and hence the desired commutator $[F \diamond \widetilde{F}, a]$. In each of the terms in $w_\lambda C w_\lambda$, we can, neglecting B-compact operators, replace F by $\widetilde{F}$ in view of the agreement condition (3.4) and freely commute F with the other factors in C_j, and so we obtain
$$w_\lambda C w_\lambda = w_\lambda \big([\psi_1, a] F \psi_1 + [\psi_2, a] \widetilde{F} \psi_2 + \psi_1 F [\psi_1, a] + \psi_2 \widetilde{F} [\psi_2, a]\big) w_\lambda$$
$$\sim w_\lambda \big([\psi_1, a] \psi_1 F + [\psi_2, a] \psi_2 F + \psi_1 [\psi_1, a] F + \psi_2 [\psi_2, a] F\big) w_\lambda$$
$$= w_\lambda [\psi_1^2 + \psi_2^2, a] F w_\lambda = 0,$$
and we are done. Thus, we have proved that $x \diamond \widetilde{x}$ is a Kasparov module.

4. Let us prove that $x \diamond \widetilde{x}$ agrees with x on J_1 and with $\widetilde{x}$ on J_2. Let $c, d \in J_1$. Then
$$c(\psi_1 F \psi_1 + \psi_2 \widetilde{F} \psi_2) d \sim c(\psi_1 F \psi_1 + \psi_2 F \psi_2) d \sim c(\psi_1^2 F + \psi_2^2 F) d = cFd.$$
The computation for J_2 and $\widetilde{x}$ is similar.

5. Finally, let us prove that $x \diamond \widetilde{x}$ is independent modulo B-compact perturbations of the choice of the partition of unity (3.2). The argument goes as follows. Let $\psi_1'^2 + \psi_2'^2 = 1$ be another partition of unity with the same properties. Then
$$C \stackrel{\text{def}}{=} \psi_1 F \psi_1 + \psi_2 \widetilde{F} \psi_2 - \psi_1' F \psi_1' - \psi_2' \widetilde{F} \psi_2'$$
$$- (\psi_1 - \psi_1') F \psi_1 + \psi_1' F (\psi_1 - \psi_1') + (\psi_2 - \psi_2') \widetilde{F} \psi_2 + \psi_2' \widetilde{F} (\psi_2 - \psi_2').$$

3.3. Superposition Principle

One can readily prove (again using functional calculus) that $\psi_1 - \psi_1' \in J_0$ and $\psi_2 - \psi_2' \in J_0$. Arguing by analogy with the above, we conclude that (3.12) holds for this new operator C, and then we can again use the agreement condition (3.4) and the commutation of F with elements of A to prove that C is B-compact.

The proof of Theorem 3.9 is complete. □

In a similar way, one defines the Kasparov module $\tilde{x} \diamond x$, which agrees with $\tilde{x}$ on J_1 and with x on J_2.

Main result. Now we are in a position to state the main result of this chapter. For a Kasparov (A,B)-module y, let $[y] \in KK(A,B)$ be the corresponding class in Kasparov's KK-theory.

Theorem 3.10. *Let A be a unital C^*-algebra, let J_1 and J_2 be two ideals in A such that $J_1 + J_2 = A$, let B be a C^*-algebra, and let x and $\tilde{x}$ be two Kasparov (A,B)-modules that agree on the ideal $J_0 = J_1 \cap J_2$. Then*

$$[x] + [\tilde{x}] = [x \diamond \tilde{x}] + [\tilde{x} \diamond x]. \tag{3.13}$$

Before proceeding to the proof, let us state a corollary of this theorem.

Corollary 3.11 (Relative index theorem for Kasparov modules). *Let the assumptions of Theorem (3.10) be satisfied. Then*

$$\mathrm{ind}(x \diamond \tilde{x}) - \mathrm{ind}(x) = \mathrm{ind}(\tilde{x}) - \mathrm{ind}(\tilde{x} \diamond x) \in K_0(B).$$

The result follows by applying the mapping $KK(A,B) \longrightarrow K_0(B) \equiv KK(\mathbb{C}, B)$ induced by the natural embedding $\mathbb{C} \longrightarrow A$ to (3.13).

Proof of Theorem 3.10. The full proof involves lengthy but routine computations, many of which resemble those carried out above in the construction of $x \diamond \tilde{x}$. So we omit the technicalities and only describe the crucial steps.

1. Under our identifications, the formula

$$\begin{pmatrix} \xi \\ \eta \end{pmatrix} \longmapsto \begin{pmatrix} \psi_1 \xi + \psi_2 \eta \\ -\psi_2 \xi + \psi_1 \eta \end{pmatrix} \tag{3.14}$$

gives a well-defined unitary isomorphism

$$U: H \oplus \tilde{H} \longrightarrow (H \diamond \tilde{H}) \oplus (\tilde{H} \diamond H) \tag{3.15}$$

of Hilbert B-modules such that

$$U(F \oplus \tilde{F})U^* \sim (F \diamond \tilde{F}) \oplus (\tilde{F} \diamond F). \tag{3.16}$$

The inverse mapping is given by

$$\begin{pmatrix} \xi \\ \eta \end{pmatrix} \longmapsto \begin{pmatrix} \psi_1 \xi - \psi_2 \eta \\ \psi_2 \xi + \psi_1 \eta \end{pmatrix}. \tag{3.17}$$

(It may be an instructive exercise to verify that, with our identifications, all these formulas are well defined indeed.)

2. The mapping (3.14)–(3.15) is an isomorphism of (A,B)-bimodules, where the action of A on $(H \diamond \tilde{H}) \oplus (\tilde{H} \diamond H)$ is the direct sum of the actions $\rho \diamond \tilde{\rho}$ and $\tilde{\rho} \diamond \rho$, while the action of A on $H \oplus \tilde{H}$ is given by the matrix

$$\hat{\rho}(a) = \begin{pmatrix} \psi_1 a \psi_1 + \psi_2 a \psi_2 & \psi_2 a \psi_1 - \psi_1 a \psi_2 \\ \psi_1 a \psi_2 - \psi_2 a \psi_1 & \psi_1 a \psi_1 + \psi_2 a \psi_2 \end{pmatrix}, \qquad a \in A. \tag{3.18}$$

(This matrix is of course obtained from the diagonal matrix $\begin{pmatrix} a & 0 \\ 0 & a \end{pmatrix}$ by conjugation with U.) Note that the right-hand side in (3.18) is well defined, because the off-diagonal entries lie in J_0 and hence take H as well as $\tilde{H}$ to H_0, which lies in both. Moreover, a straightforward computation shows that $[\hat{\rho}(a), F \oplus \tilde{F}] \sim 0$, because F and $\tilde{F}$ agree on J_0.

3. Thus, it remains to prove that the Kasparov modules $(H \oplus \tilde{H}, \rho \oplus \tilde{\rho}, F \oplus \tilde{F})$ and $(H \oplus \tilde{H}, \hat{\rho}, F \oplus \tilde{F})$ define the same element in $KK(A,B)$. To this end, we construct a homotopy $(H \oplus \tilde{H}, \hat{\rho}_t, F \oplus \tilde{F})$ of Kasparov modules such that $\hat{\rho}_0 = \rho \oplus \tilde{\rho}$ and $\hat{\rho}_1 = \hat{\rho}$. Namely, for $a \in A$ we set

$$\hat{\rho}_t(a) = \begin{pmatrix} \psi_{1t} a \psi_{1t} + \psi_{2t} a \psi_{2t} & \psi_{2t} a \psi_{1t} - \psi_{1t} a \psi_{2t} \\ \psi_{1t} a \psi_{2t} - \psi_{2t} a \psi_{1t} & \psi_{1t} a \psi_{1t} + \psi_{2t} a \psi_{2t} \end{pmatrix}, \tag{3.19}$$

where $\psi_{1t} = t \psi_1$ and $\psi_{2t} = \sqrt{1 - t^2 \psi_1^2}$. Let us show that formula (3.19) indeed gives a well-defined homotopy between $\rho \oplus \tilde{\rho}$ and $\hat{\rho}$.

First, we should prove that the operator (3.19) is well defined on $H \oplus \tilde{H}$. To this end, it suffices to show that the off-diagonal entries lie in J_0. We have

$$\psi_{2t} = (1-t^2)^{1/2} + \left((1-t^2+t^2\psi_2^2)^{1/2} - (1-t^2)^{1/2} \right).$$

The term in parentheses lies in J_2 by functional calculus, and we obtain

$$\psi_{2t} a \psi_{1t} - \psi_{1t} a \psi_{2t} \equiv t\sqrt{1-t^2}[a, \psi_1] \mod J_0.$$

But we already know that $[a, \psi_1] \in J_0$, and so we are done.

Now the properties

$$\psi_{1t}^2 + \psi_{2t}^2 = 1, \qquad [\psi_{1t}, \psi_{2t}] = 0$$

imply, after straightforward computations, that $\hat{\rho}_t$ is indeed a homomorphism and that

$$[\hat{\rho}_t(a), F \oplus \tilde{F}] \sim 0.$$

(When proving this, one should again use the fact that F and $\tilde{F}$ agree on J_0.)

The proof of Theorem 3.10 is complete. □

Part II

Examples

Chapter 4

Elliptic Operators on Noncompact Manifolds

In this chapter, we discuss various relative index theorems of Gromov–Lawson type, which have important applications in geometry and topology. We, however, do not dwell on these applications (for which we refer the reader to the literature) but focus our attention on the relationship between these theorems and the general relative index theorems given in Part I of the book.

4.1 Gromov–Lawson Theorem

Apparently, the first theorems of this type were stated in Gromov and Lawson's seminal paper [32]. Let us briefly recall these theorems. The setting in which they are stated is as follows. Let X be a Riemannian manifold, and let S be a vector bundle over X that is a bundle of modules over the Clifford bundle of X. The generalized Dirac operator $D \colon \Gamma(S) \longrightarrow \Gamma(S)$ acting on sections of the bundle S is locally given by the formula

$$D = \sum_j e_j . \nabla_{e_j},$$

where e_j is a local orthonormal frame on X, ∇_{e_j} is the covariant differentiation, and the dot stands for Clifford multiplication. If X is oriented and even-dimensional (which we will assume), then S splits, $S = S^+ \oplus S^-$ (the splitting is parallel and orthogonal), so that the operator D can be represented in the block matrix form

$$D = \begin{pmatrix} 0 & D^- \\ D^+ & 0 \end{pmatrix}, \qquad D^+ \colon \Gamma(S^+) \longrightarrow \Gamma(S^-), \qquad D^- \colon \Gamma(S^-) \longrightarrow \Gamma(S^+).$$

Now suppose that X is a complete Riemannian manifold; in other words, X is complete as a metric space, or, equivalently, every geodesic in X is infinitely extendable. Then D

with domain consisting of compactly supported sections of S is essentially self-adjoint in $L^2(X;S)$. We denote the closure of D by the same letter. Then $D^- = (D^+)^*$. If D is an (unbounded) Fredholm operator in $L^2(X;S)$, then so are $D^\pm$, and we write $\operatorname{ind} D$ meaning $\operatorname{ind} D^+ = \dim \ker D^+ - \dim \ker D^-$. There is a simple sufficient condition for D to be Fredholm. The Weitzenböck formula says that

$$D^*D = \nabla^*\nabla + \mathscr{R},$$

where $\mathscr{R}$ is a symmetric bundle endomorphism that can be computed explicitly (e.g., see [32, Proposition 2.5]). For example, $\mathscr{R} = \varkappa/4$, $\varkappa$ being the scalar curvature of the metric, for the Dirac operator acting on the sections of the complex bundle of spinors on an even-dimensional spin manifold X. If D is *positive at infinity* in the sense that[1] $\mathscr{R} \geq cI$, $c > 0$, outside a compact subset of X, then D is Fredholm.

Now let $D_j \colon \Gamma(S_j) \longrightarrow \Gamma(S_j)$, $j = 0, 1$, be two generalized Dirac operators on complete oriented even-dimensional Riemannian manifolds X_j, and suppose that these operators coincide outside some compact sets $K_j \subset X_j$. More precisely, this means that the sets $X_0 \setminus K_0$ and $X_1 \setminus K_1$ are (isometrically) diffeomorphic (in what follows, we identify them, $X_0 \setminus K_0 \simeq X_1 \setminus K_1 \stackrel{\text{def}}{=} \Omega$, see Fig. 4.1), the restrictions of the bundles S_j to Ω are isomorphic (in what follows, we identify them as well), and, as a consequence, the Dirac operators D_0 and D_1 coincide on Ω. In this situation, Gromov and Lawson define the *"topological"* rel-

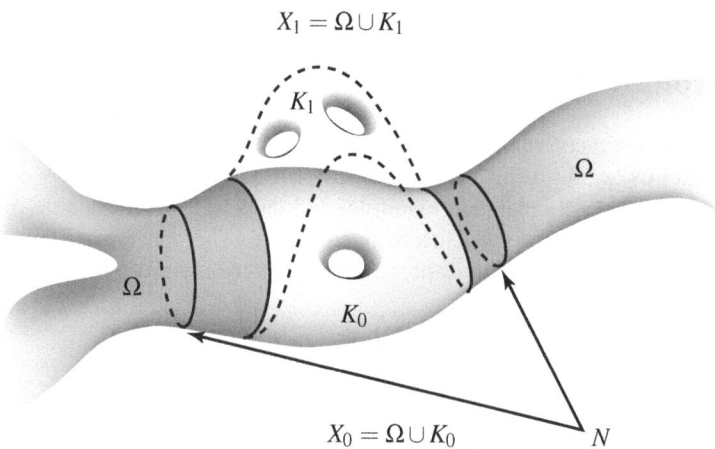

Figure 4.1: The manifolds X_0 and X_1

ative index of D_1 and D_2 as follows: they compactify the manifolds X_j by cutting away

---

[1] Here I is the identity automorphism of the bundle S and the inequality is understood in the sense of the corresponding quadratic forms on the fibers.

4.1. Gromov–Lawson Theorem

a noncompact part of Ω along some smooth compact hypersurface $N \subset \Omega$ and by pasting some compact part instead, thus obtaining compact Riemannian manifolds $\widehat{X}_j$ (see Fig. 4.2), extend[2] the bundles S_j to bundles of Clifford modules over X_j, and consider the corresponding generalized Dirac operators $\widehat{D}_j$. Then they set

$$\operatorname{ind}_t(D_1, D_0) = \operatorname{ind} \widehat{D}_1 - \operatorname{ind} \widehat{D}_0. \tag{4.1}$$

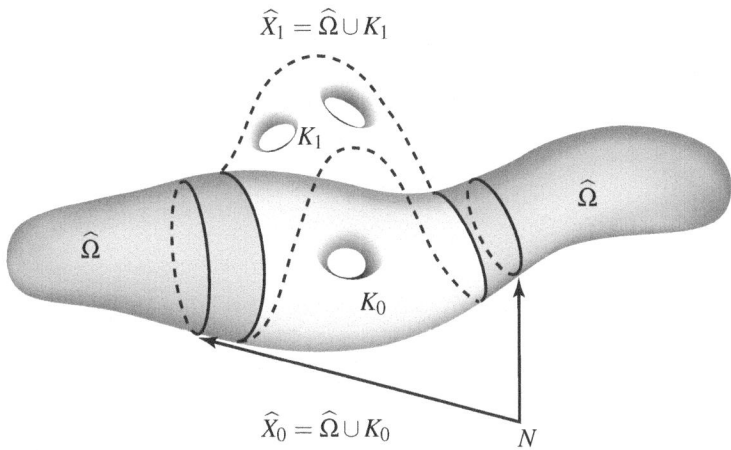

Figure 4.2: The manifolds $\widehat{X}_0$ and $\widehat{X}_1$

Then the main assertions are as follows:

1. The topological relative index is well defined, that is, independent of the specific choice of the compactification and the continuations of the bundles and Dirac operators.

2. If the original operators D_0 and D_1 are positive at infinity (and hence Fredholm), then the "*analytical*" relative index

$$\operatorname{ind}_a(D_1, D_0) = \operatorname{ind} D_1 - \operatorname{ind} D_0 \tag{4.2}$$

coincides with the topological one,

$$\operatorname{ind}_a(D_1, D_0) = \operatorname{ind}_t(D_1, D_0). \tag{4.3}$$

---

[2] One can always choose the compact part to be pasted in such a way that the extension is possible.

The latter assertion is the Gromov–Lawson relative index theorem [32, Theorem 4.18].

It was further indicated in [32] (and later in [16, Theorem 10.2]) that the topological relative index (4.1) can also be defined as follows. Cut away the infinite parts of X_0 and X_1 along N and paste the remaining compact parts $\widetilde{X}_0$ and $\widetilde{X}_1$ together[3] to obtain the compact Riemannian manifold[4] $\widetilde{X}_1 \cup (-\widetilde{X}_0)$ (see Fig. 4.3). The restrictions of the operators D_j to $\widetilde{X}_j$

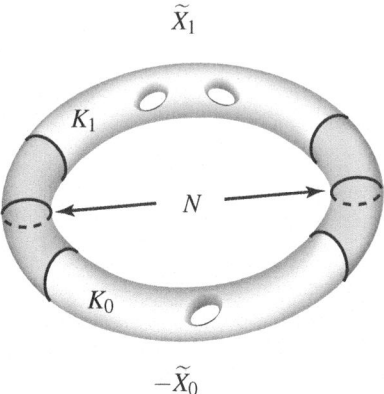

Figure 4.3: The manifold $\widetilde{X}_1 \cup (-\widetilde{X}_0)$

naturally agree on N and can be pasted together to form a generalized Dirac operator D on $\widetilde{X}_1 \cup (-\widetilde{X}_0)$. Then

3. One has

$$\operatorname{ind}_t(D_1, D_0) = \operatorname{ind} D. \tag{4.4}$$

Finally, the paper [32] contains the following generalization of the relative index theorem. Suppose that the sets K_0 and K_1 outside which the operators D_0 and D_1 coincide may be noncompact but still there exists a hypersurface $N \subset \Omega$ that cuts away an infinite part of Ω (see Fig. 4.4). Furthermore, let D_0 and D_1 be positive at infinity.[5] Then we can, as before, cut away the infinite part of Ω along N from both X_0 and X_1, thus obtaining (noncompact) Riemannian manifolds $\widetilde{X}_0$ and $\widetilde{X}_1$; we can then paste these manifolds together along N, which gives the noncompact Riemannian manifold $\widetilde{X}_1 \cup (-\widetilde{X}_0)$ (Fig. 4.5). This manifold is complete, and the operators D_0 and D_1 restricted to $\widetilde{X}_0$ and $\widetilde{X}_1$ naturally paste together to form a generalized Dirac operator D on $\widetilde{X}_1 \cup (-\widetilde{X}_0)$. This operator is positive at infinity and hence has a well-defined index. Thus, we set

$$\operatorname{ind}_t(D_1, D_0) \stackrel{\text{def}}{=} \operatorname{ind} D. \tag{4.5}$$

---

[3] For this procedure to be done, one should first deform the metric in a collar neighborhood of N to a product form.

[4] The minus sign accounts for the orientation reversal.

[5] This is clearly an overkill; to define the relative index, it suffices to require that D_0 and D_1 be strictly positive at infinity only in K_0 and K_1, respectively.

4.1. Gromov–Lawson Theorem

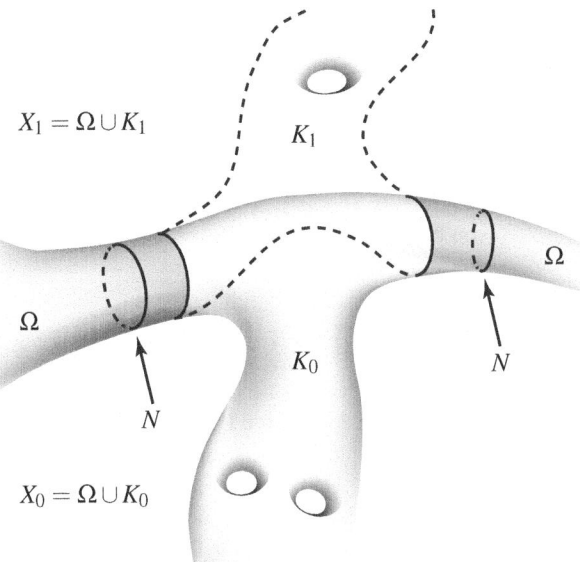

Figure 4.4: The manifolds X_0 and X_1

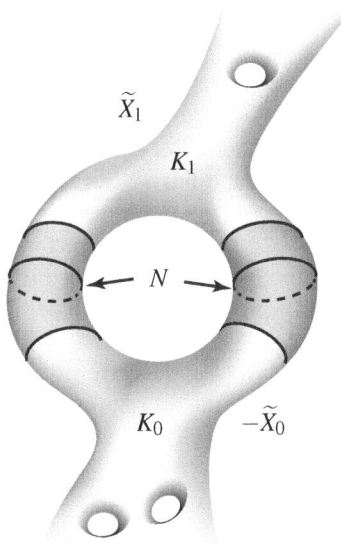

Figure 4.5: The manifold $\widetilde{X}_1 \cup (-\widetilde{X}_0)$ (the noncompact case)

(This agrees with (4.4) if K_0 and K_1 are actually compact.) The generalized Gromov–Lawson relative index theorem states that

4. In this situation,
$$\operatorname{ind}_a(D_1, D_0) = \operatorname{ind}_t(D_1, D_0). \tag{4.6}$$

In fact, each of the statements 1–4 is a special case of the general relative index superposition principle given by Theorem 0.10 (or, more precisely, can be reduced to it).

4.2 Bunke Theorem

Bunke [22] developed a K-theoretic version of the Gromov–Lawson relative index theorem. For a $\mathbb{Z}_2$-graded C^*-algebra B, he considered Dirac operators D acting on sections of bundles E of projective graded Hilbert B-modules (see [47] or [13]) over complete Riemannian manifolds M. Let $C_g(M) \subset C(M)$ be the C^*-subalgebra defined as the closure of the set of infinitely differentiable functions f on M such that $\|df\| \in C_0(M)$. If the Dirac operator D is *invertible at infinity* in the sense that there exists a nonnegative function $f \in C_0^\infty(M)$ such that the operator $D^2 + f \colon H^2(M, E) \longrightarrow H^0(M, E)$, where $H^j(M, E)$ is the Sobolev space of sections of the bundle E (see [47]), is boundedly invertible, then the operator
$$F = D(D^2 + f)^{-1/2} \tag{4.7}$$
specifies a well-defined Kasparov $(C_g(M), B)$-module and hence a class in $KK(C_g(M), B)$, which we, following [22], denote by $[M] \in KK(C_g(M), B)$. (Thus, the bundle E and all additional structures needed to define the Dirac operator get absorbed in this notation.)

Now consider the following situation. Let D_1 and D_2 be two Dirac operators of this kind on two complete Riemannian manifolds M_1 and M_2, and assume that there exist isomorphic compact hypersurfaces $N_j \subset M_j$ such that N_j divides M_j into two parts W_j and V_j (that is, $M_j = W_j \cup_{N_j} V_j$), $j = 1, 2$, and all structures on the two manifolds are isomorphic in some tubular neighborhoods U_j of N_j (see Fig. 4.6).

Then we can make the following surgery: we cut M_1 and M_2 along $N_1 \simeq N_2 = N$ and interchange the pieces, i.e., glue V_2 to W_1 and V_1 to W_2 along N, thus obtaining two new manifolds M_3 and M_4 (see Fig. 4.7) and the corresponding Dirac operators.

Assume that D_1 and D_2 are invertible at infinity. Then one can prove that so are D_3 and D_4, and we obtain four KK-theory elements
$$[M_j] \in KK(C_g(M_j), B), \qquad j = 1, \ldots, 4.$$

Note that each $C_g(M_j)$ is a unital C^*-algebra, and hence we have the embedding $i_j \colon k \longrightarrow C_g(M_j)$ (where k is the field $\mathbb{R}$ or $\mathbb{C}$), which induces the corresponding index mapping
$$i_j^* \colon KK(C_g(M_j), B) \longrightarrow KK(k, B).$$

Set $\{M_j\} = i_j^*[M_j]$, $j = 1, \ldots, 4$. One has the following theorem.

4.2. Bunke Theorem

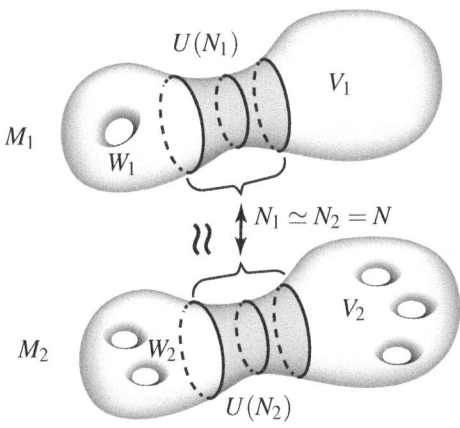

Figure 4.6: The manifolds M_1 and M_2

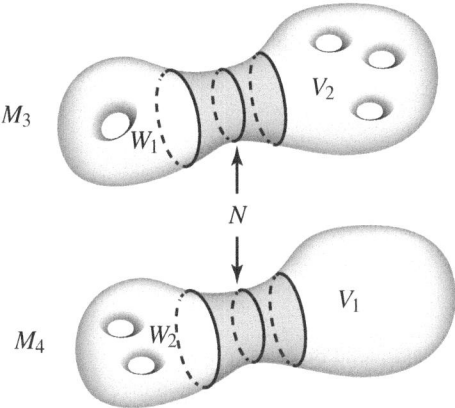

Figure 4.7: After the surgery: the manifolds M_3 and M_4

Theorem 4.1 ([22, Theorem 1.2]).

$$\{M_1\} + \{M_2\} - \{M_3\} + \{M_4\} = 0. \tag{4.8}$$

(Of course, one returns to the "classical" Gromov–Lawson theorem if $k = B = \mathbb{C}$.)

Our Theorem 3.10 is very close in spirit to Bunke's Theorem 4.1 and can be viewed as a natural generalization of the latter to the case of nontrivial (and possibly noncommutative) algebras A. Let us show how Theorem 3.10 applies to the situation described above and how Bunke's theorem can be derived from it. Let $U \supseteq U(N_1) \simeq U(N_2)$ be a common part of our manifolds M_1 and M_2 (and hence of M_3 and M_4).[6] Next, let $\widetilde{U} \subset U$ be a smaller tubular neighborhood of N (see Fig. 4.8).

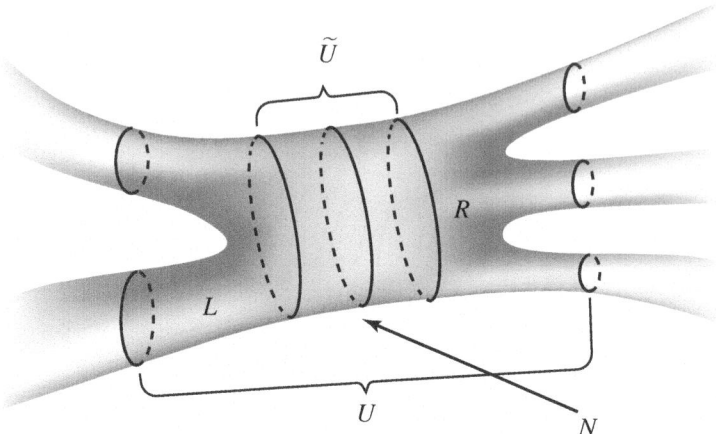

Figure 4.8: Common part of all four manifolds M_j, $j = 1, \ldots, 4$

Thus, we have
$$U = L \sqcup \widetilde{U} \sqcup R.$$

Now let us introduce the following unital algebra $A \subset C(\overline{U})$:

$$A = \{f \in C(U) \colon \forall j = 1, \ldots, 4 \,\exists f_j \in C(M) \\ \text{such that } f_j|_U = f \text{ and } f_j \text{ is locally constant on } M_j \setminus U\}. \quad (4.9)$$

(In particular, each $f \in A$ is constant on every connected component of ∂U.) The functions f_j in (4.9) are uniquely determined and belong to the respective algebras $C_g(M_j)$; thus, we have a natural embedding $\mu_j \colon A \longrightarrow C_g(M_j)$ and the associated mapping

$$\mu_j^* \colon KK(C_g(M_j), B) \longrightarrow KK(A, B) \quad (4.10)$$

for each $j = 1, \ldots, 4$. Set

$$[[M_j]] = \mu_j^*[M_j] \in KK(A, B), \quad j = 1, \ldots, 4. \quad (4.11)$$

These classes correspond to Kasparov (A, B)-modules F_j constructed from the Dirac operators D_j by formula (4.7). Now we can apply Theorem 3.10. Indeed, consider the following ideals of the algebra A:

$$J_1 = \{f \in A \colon f|_R = 0\}, \quad J_2 = \{f \in A \colon f|_L = 0\}.$$

---

[6] Actually, the common part U (i.e., a part to which all relevant isomorphisms can be continued) may well be larger than the tubular neighborhood $U(N_1) \simeq U(N_2)$ of N; our argument remains valid in this case.

Then $J_1 + J_2 = A$, $J_1 \cap J_2 = C_0(\widetilde{U})$, and one can readily see that

$$F_3 = F_1 \diamond F_2, \qquad F_4 = F_2 \diamond F_1.$$

We obtain the following theorem.

Theorem 4.2. *One has*

$$[[M_1]] + [[M_2]] - [[M_3]] - [[M_4]] = 0 \in KK(A,B).$$

Now Bunke's relative index Theorem 4.1 follows by Corollary 3.11 for $k = \mathbb{C}$ and by an obvious analog of this corollary for $k = \mathbb{R}$.

4.3 Roe's Relative Index Construction

In a somewhat different way the relative index construction due to Gromov and Lawson was put in context of operator algebras and their K-theory and K-homology by Roe [70–72] and Roe–Siegel [73]). They used sheaves of C^*-algebras, Paschke duality and coarse geometry to define the relative homology class of a pair of elliptic operators in the following setting.

Let X be a locally compact metrizable set, and let $Z \subset X$ be a closed subset. Next, let M_1 and M_2 be two manifolds bearing elliptic first-order symmetric differential operators D_1 and D_2. Next, let

$$c_j \colon M_j \to X, \qquad j = 1, 2,$$

be continuous proper maps (referred to as the *control maps*). Suppose that the operators D_1 and D_2 coincide outside the sets $Z_j = c_j^{-1}(Z)$. (This means that there is a diffeomorphism between $M_1 \setminus Z_1$ and $M_2 \setminus Z_2$ covered by a bundle isomorphism such that the conjugation with these isomorphisms takes one operator to the other.) This set of *relative elliptic data* specifies a well-defined K-homology class in $K_*(Z)$, and Proposition 4.8 in [73] states that this homology class depends only on the behavior of the operators in question "in a neighborhood of Z." This is a generalization of the Gromov–Lawson theorem (which is obtained if $X = \mathbb{R}_+$ and $Z = \{0\}$).

Note that this result pertains only to the commutative case. It would be of interest to obtain a similar theorem in the noncommutative case, thus supplementing the results in Chapter 2 of the present book. In general, if two Fredholm modules F_1 and F_2 over a C^*-algebra A coincide on an ideal $J \subset A$, then the difference $[F_1] - [F_2] \in K^*(A)$ lies in the image of the natural mapping

$$\pi^* \colon K^*(A/J) \longrightarrow K^*(A)$$

but $(\pi^*)^{-1}([F_1] - [F_2])$ is not uniquely determined, and additional information is needed to define an element in $K^*(A/J)$ corresponding to the relative elliptic data.

Chapter 5

Applications to Boundary Value Problems

5.1 Preliminaries

Here we introduce some notation and present technical tools that come in handy when applying the general superposition principle for the relative index to the theory of boundary value problems for elliptic differential operators.

5.1.1 Notation

Let X be a smooth compact n-dimensional manifold with boundary $\partial X = Y$ that is a smooth closed manifold of dimensional $n-1$. We choose and fix a representation of some collar neighborhood U of the boundary in the form of a direct product

$$U \simeq Y \times [0,1), \tag{5.1}$$

where Y is taken to $Y \times \{0\}$ by the identity mapping. The coordinate on $[0,1)$ will be denoted by t, and local coordinates on the boundary by $y = (y_1, \ldots, y_{n-1})$, so that local coordinates on X in U have the form

$$x = (x_1, \ldots, x_n) = (y,t).$$

If E is a vector bundle over X, then the restriction $E|_U$ is isomorphic to the lift to U of the restriction $E|_Y$ of the same bundle to the boundary:

$$E_U \simeq \pi_U^* E|_Y, \tag{5.2}$$

where $\pi_U : U \to Y$ is the projection naturally associated with the representation (5.1).
Now let

$$\widehat{D} : C^\infty(X, E_1) \to C^\infty(X, E_2) \tag{5.3}$$

be an elliptic differential operator of order m on X acting in sections of finite-dimensional vector bundles E_1 and E_2. Then, using the trivialization (5.1) and the associated representations (5.2) of E_1 and E_2 over U as the lifts of $E_1|_Y$ and $E_2|_Y$, we can represent the operator (5.3) in U in the form

$$\widehat{D} = \sum_{j=0}^{m} \widehat{D}_j(t) \left(-i\frac{\partial}{\partial t}\right)^j, \tag{5.4}$$

where

$$\widehat{D}_j(t) : C^\infty(Y, E_1|_Y) \to C^\infty(Y, E_2|_Y)$$

is a differential operator of order $m - j$ depending on the parameter t and acting in sections of bundles over Y. Next, the coefficient $\widehat{D}_m(t)$ is a differential operator of order 0, i.e., a bundle homomorphism, and since $\widehat{D}$ is elliptic, this coefficient is a bundle isomorphism. Dividing the operator $\widehat{D}$ in U by this coefficient on the left, we can assume without loss of generality that the bundles $E_1|_Y$ and $E_2|_Y$ coincide and the coefficient itself is the identity operator.

The operator family

$$\mathfrak{D}(p) = \sum_{j=0}^{m} \widehat{D}_j(0) p^j : H^s(Y) \to H^{s-m}(Y) \tag{5.5}$$

acting in Sobolev spaces[1] on Y and obtained from the representation (5.4) by freezing the coefficients at the boundary $t = 0$ and by replacing the operator $-i\partial/\partial t$ with the variable p will be called the *conormal symbol* of the operator $\widehat{D}$.

If $u \in H^s(X)$ is an element of a Sobolev space on X, then for $s > m - 1/2$ by trace theorems we have a well-defined jet of order $m - 1$ of u on Y. With regard to the identifications (5.1) and (5.2), it can be rewritten in the form

$$j_X^{m-1} u = \left(u\Big|_{t=0}, \frac{\partial u}{\partial t}\Big|_{t=0}, \ldots, \frac{\partial^{m-1} u}{\partial t^{m-1}}\Big|_{t=0} \right) \in H^{s-1/2}(Y) \oplus \cdots \oplus H^{s-m+1/2}(Y). \tag{5.6}$$

Boundary value problems for $\widehat{D}$ are stated in terms of the boundary jet (5.6) of u, to which one applies some differential or pseudodifferential operators. Since for $m > 1$ the space on the right-hand side in (5.6), which for brevity will be denoted by

$$\mathcal{H}_m^{s-1/2}(Y) = \bigoplus_{k=0}^{m-1} H^{s-k-1/2}(Y), \tag{5.7}$$

is a direct sum of Sobolev spaces of *various orders*, we see that the orders of pseudodifferential operators in such spaces must be understood in the sense of Douglis–Nirenberg.

---

[1] In what follows, we usually omit the bundles in the notation of Sobolev spaces.

5.1.2 General boundary value problem

General boundary value problems, which include classical boundary value problems as well as nonlocal problems of the Atiyah–Patodi–Singer type (in particular, nonhomogeneous) were introduced in [76]. Let $\widehat{D}$ be an elliptic differential operator (5.3) on a manifold X. A *general boundary value problem* for $\widehat{D}$ is a problem of the form

$$\begin{cases} \widehat{D}u &= f \in H^{s-m}(X), \\ \widehat{B}j_Y^{m-1}u &= g \in \mathscr{L}, \end{cases} \qquad (5.8)$$

where $s > m - 1/2$, the element $u \in H^s(X)$ is to be found, $\mathscr{L}$ is a Hilbert space, and $\widehat{B}$ is a continuous linear operator in the spaces

$$\widehat{B} : \mathscr{H}_m^{s-1/2}(Y) \to \mathscr{L}. \qquad (5.9)$$

Ordinary boundary value problems are the special case in which $\mathscr{L}$ is a Sobolev space of sections of some vector bundle over the boundary and B is a (pseudo)differential operator. If $\widehat{D}$ is the Dirac operator on an even-dimensional manifold X, $\mathscr{L}$ is the positive spectral subspace of the tangential Dirac operator, and B is the orthogonal projection onto $\mathscr{L}$, then we arrive at the Atiyah–Patodi–Singer problem [6], or, more precisely, at a more general problem in which the nonlocal boundary data may be nonzero.

As shown by these examples, of main interest is the case in which $\mathscr{L}$ is not an abstract Hilbert space but rather a subspace of some Sobolev space on the boundary[2] and B is a pseudodifferential operator. More precisely, we shall consider only subspaces that are ranges of pseudodifferential projections. If $\widehat{P}$ is a pseudodifferential projection onto some subspace $\widehat{L}$ of a Sobolev space of sections of some vector bundle F over Y, then the principal symbol $P = \sigma(\widehat{P})$ is a projection onto a subbundle $L \subset \pi^*F$ over T_0^*Y, where $\pi : T_0^*Y \to Y$ is the natural projection. The subbundle L is called the *principal symbol* of $\widehat{L}$. The pseudodifferential version of the general boundary value problem (5.8) for an unknown function $u \in H^s(X, E_1)$ has the form

$$\begin{cases} \widehat{D}u &= f \in H^{s-m}(X, E_2), \\ \widehat{B}j_Y^{m-1}u &= g \in \widehat{P}H(Y, F), \end{cases} \qquad (5.10)$$

where $H(Y, F)$ is a Sobolev space of sections of a bundle F over the boundary (we intentionally omit the index on this space, since it can be a usual Sobolev space or a space of the form $\mathscr{H}_m^s(Y)$) and $\widehat{B} : \mathscr{H}_m^{s-1/2} \to H(Y, F)$ is a pseudodifferential operator such that $R(\widehat{B}) \subseteq R(\widehat{P})$. (The last inclusion necessarily implies that $R(B) \subseteq R(P) = L$.)

The general boundary value problems (5.8) include problems that are a straightforward (inhomogeneous) analog of the Atiyah–Patodi–Singer problem. Namely, let an operator $\widehat{D}$ of order m be given. On the basis of the conormal symbol (5.5) of $\widehat{D}$, let us construct a pseudodifferential projection

$$\widehat{P}_+ : \mathscr{H}_m^s(Y) \to \mathscr{H}_m^s(Y) \qquad (5.11)$$

---

[2] In particular, the entire Sobolev space.

in the Cauchy data space (5.7). The construction is as follows (see [54]). On the basis of the operator family (5.5), we construct the matrix operator

$$\mathfrak{A} = \begin{pmatrix} 0 & 1 & 0 & \cdots & 0 \\ 0 & 0 & 1 & \cdots & 0 \\ 0 & 0 & 0 & \cdots & 1 \\ -\widehat{D}_0 & -\widehat{D}_1 & -\widehat{D}_2 & \cdots & -\widehat{D}_{n-1} \end{pmatrix} : \mathscr{H}_m^s(Y) \to \mathscr{H}_m^{s-1}(Y),$$

where $\widehat{D}_j \stackrel{\text{def}}{=} \widehat{D}_j(0)$, in the Cauchy data space. The operator $p - \mathfrak{A}$, $p \in \mathbb{C}$, is invertible if and only if so is the operator $\mathfrak{D}(p)$, and moreover,

$$(p - \mathfrak{A})^{-1} = \mathfrak{D}(p)^{-1} \mathfrak{Q}(p),$$

where the entries $\widehat{q}_{jk}(p)$, $j,k = 0, \ldots, m-1$ of the matrix $\mathfrak{Q}(p)$ are differential operators polynomially depending on p whose total order (with regard to the parameter, to which we assign the unit weight) does not exceed

$$\operatorname{ord} \widehat{q}_{jk}(p) \leq m - 1 + j - k.$$

Since the operator $\widehat{D}$ is elliptic, it follows that the polynomial family $\mathfrak{D}(p)$ is elliptic with parameter p [79] in the double sector

$$\Lambda_\varepsilon = \{|\arg p| < \varepsilon\} \cup \{|\pi - \arg p| < \varepsilon\}$$

on the complex p-plane for some $\varepsilon > 0$ and elliptic in the usual sense for all $p \in \mathbb{C}$. It follows that the family $\mathfrak{D}(p)$ is finitely meromorphically invertible in the entire complex plane and the sector Λ_ε contains only finitely many poles of the operator function $\mathfrak{D}^{-1}(p)$. Thus, the operator $\mathfrak{D}(p)$ (and hence $p - \mathfrak{A}$) is invertible on the line $\operatorname{Im} p = \delta$ for all sufficiently small $\delta > 0$.

Set

$$\widehat{P}_+ = -\frac{\widehat{\mathfrak{A}}}{2\pi i} \int_{-\infty + i\delta}^{+\infty + i\delta} (p - \mathfrak{A})^{-1} \frac{dp}{p}, \tag{5.12}$$

where $\delta > 0$ is sufficiently small. This integral specifies a well-defined continuous projection

$$\widehat{P}_+ : \mathscr{H}_m^s(Y) \to \mathscr{H}_m^s(Y) \tag{5.13}$$

in the Cauchy data space (e.g., see [54]). This projection corresponds to the spectral points of the operator $\mathfrak{A}$ in the upper half-plane. If $\widehat{D}$ is the Dirac operator, then the projection $\widehat{P}_+$ thus introduced coincides with the Atiyah–Patodi–Singer spectral projection. In the following, we also set

$$\widehat{P}_- \stackrel{\text{def}}{=} 1 - \widehat{P}_+. \tag{5.14}$$

The *spectral boundary value problem* is problem (5.8) of the special form

$$\begin{cases} \widehat{D}u = f \in \mathscr{H}^{s-u}(X), \\ \widehat{P}_+ j_Y^{m-1} u = g \in \widehat{P}_+ \mathscr{H}_m^{s-1/2}(Y). \end{cases} \tag{5.15}$$

5.1. Preliminaries

Problem (5.15), which will be denoted by $(\widehat{D}, \widehat{P}_+)$, is always Fredholm. The index of a general Fredholm boundary value problem (5.8), which will be denoted by $(\widehat{D}, \widehat{B})$, is expressed by the formula

$$\operatorname{ind}(\widehat{D}, \widehat{P}) = \operatorname{ind}(\widehat{D}, \widehat{P}_+) + \operatorname{ind}(\widehat{B} : \widehat{P}_+ \mathscr{H}_m^{s-1/2}(Y) \to \mathscr{L}). \tag{5.16}$$

Problem (5.10) is Fredholm if and only if the principal symbol B of the operator $\widehat{B}$ is an isomorphism between the principal symbol L_+ of the subspace

$$\widehat{L}_+ = \widehat{P}_+ \mathscr{H}_m^{s-1/2}(Y)$$

and L. In this case, the above general index formula (5.16) holds.

5.1.3 Model boundary value problems on the cylinder

Applications of the superposition principle and the corresponding surgery to boundary value problems, given later on in this chapter, use model boundary value problems on the cylinder as the simplest model to which more general problems are reduced by surgery. In this subsection, we consider these model problems.

Let Y be a closed C^∞ manifold. On the cylinder

$$C = Y \times [-1, 1] \tag{5.17}$$

with boundary

$$\partial C = (Y \times \{-1\}) \cup (Y \times \{+1\})$$

consisting of two separate components (faces) $Y \times \{\pm 1\}$, consider an elliptic differential operator D of order m with coefficients independent of the coordinate $t \in [-1, 1]$,

$$\widehat{D} = \left(-i\frac{\partial}{\partial t}\right)^m + \sum_{j=0}^{m-1} \widehat{D}_j \left(-i\frac{\partial}{\partial t}\right)^j. \tag{5.18}$$

Here $\widehat{D}_j$ is a differential operator of order $m - j$ on Y; in accordance with the preceding, we assume that $\widehat{D}_m$ (the coefficient of $(-i\partial/\partial t)^m$) is the identity operator.

The conormal symbol of $\widehat{D}$ on each of the faces has the form

$$\mathfrak{D}_{-1}(p) = p^m + \sum_{j=0}^{m-1} \widehat{D}_j p^j \quad \text{on} \quad Y \times \{-1\}, \tag{5.19}$$

$$\mathfrak{D}_1(p) - \mathfrak{D}_{-1}(-p) \equiv (-p)^m + \sum_{j=0}^{m-1} \widehat{D}_j(-p)^j \quad \text{on} \quad Y \times \{1\}. \tag{5.20}$$

We denote $\mathfrak{D}_{-1}(p)$ simply by $\mathfrak{D}(p)$ and the corresponding positive spectral projection in $\mathscr{H}_m^s(Y)$ by $\widehat{P}_+$. Then the positive spectral projection corresponding to $\mathfrak{D}_1(p)$ differs from

$\widehat{P}_- = 1 - \widehat{P}_+$ by a finite-dimensional operator (and coincides with $\widehat{P}_-$ if $D(p)$ is invertible for all $p \in \mathbb{R}$).

Model problem 1 (a spectral problem)

$$\begin{cases} \widehat{D}u = f \in H^s(C), \\ \widehat{P}_+ j^{m-1}_{Y \times \{-1\}} u = g \in \widehat{P}_+ \mathscr{H}_m^{s-1/2}(Y), \\ \widehat{P}_- j^{m-1}_{Y \times \{1\}} u = h \in \widehat{P}_- \mathscr{H}_m^{s-1/2}(Y). \end{cases} \quad (5.21)$$

In this problem, the boundary conditions are determined by complementary projections $(\widehat{P}_+ + \widehat{P}_- = 1)$ on the faces of the cylinder.

Theorem 5.1. *The index of the model problem* (5.21) *is zero.*

Model problem 1' (a spectral problem)

$$\begin{cases} \widehat{D}u = f \in H^s(C), \\ \widehat{P}_{-1} j^{m-1}_{Y \times \{-1\}} u = g \in \widehat{P}_{-1} \mathscr{H}_m^{s-1/2}(Y), \\ \widehat{P}_1 j^{m-1}_{Y \times \{+1\}} u = h \in \widehat{P}_1 \mathscr{H}_m^{s-1/2}(Y). \end{cases} \quad (5.22)$$

Here $\widehat{P}_{-1}$ and $\widehat{P}_1$ are arbitrary pseudodifferential projections with the same principal symbols as $\widehat{P}_+$ and $\widehat{P}_-$, respectively (say, the Calderón projections).

Theorem 5.2. *The index of the model problem* (5.21) *is equal to*

$$\mathrm{ind}(\widehat{D}, \widehat{P}_{-1}, \widehat{P}_1) = \mathrm{ind}(\widehat{P}_+, \widehat{P}_{-1}) + \mathrm{ind}(\widehat{P}_-, \widehat{P}_1), \quad (5.23)$$

where $\mathrm{ind}(\widehat{P}, \widehat{Q})$ *is the relative index of two projections* $\widehat{P}$ *and* $\widehat{Q}$ *differing by a compact operator.*

Proof. This is a special case of the general formula (5.16). □

Remark 5.3. Recall that the relative index of two projections $\widehat{P}$ and $\widehat{Q}$ differing by a compact operator is defined as

$$\mathrm{ind}(\widehat{P}, \widehat{Q}) = \mathrm{ind}\{P \colon \mathrm{Im}\,\widehat{Q} \longrightarrow \mathrm{Im}\,\widehat{P}\},$$

where $\mathrm{Im}\,A$ is the range of an operator A.

Model problem 2 (a classical problem)

$$\begin{cases} \widehat{D}u = f \in H^s(C), \\ \widehat{B}_{-1} j^{m-1}_{Y \times \{-1\}} u = g \in H^k(Y, F_{-1}), \\ \widehat{B}_1 j^{m-1}_{Y \times \{1\}} u = h \in H^l(Y, F_1), \end{cases} \quad (5.24)$$

where $\widehat{B}_{-1}$ and $\widehat{B}_1$ are operators of classical boundary conditions on the faces of the cylinder satisfying the Shapiro–Lopatinskii conditions and F_{-1} and F_1 are some bundles over Y.

Theorem 5.4. *The index of the model problem 2 is equal to*

$$\operatorname{ind}(\widehat{D}, \widehat{B}_{-1}, \widehat{B}_1) = \operatorname{ind} \widehat{\mathscr{B}}, \tag{5.25}$$

where $\widehat{\mathscr{B}}$ is an elliptic operator on Y in the spaces

$$\widehat{\mathscr{B}} : \mathscr{H}_m^{s-1/2}(Y) \to H^k(Y, F_{-1}) \oplus H^l(Y, F_1) \tag{5.26}$$

with principal symbol

$$\mathscr{B} = \begin{matrix} B_1 & L_+ \\ \oplus & : \oplus \\ B_2 & L_- \end{matrix} \longrightarrow \begin{matrix} \pi^* F_{-1} \\ \oplus \\ \pi^* F_1 \end{matrix} \tag{5.27}$$

Here L_+ and L_- are the ranges of the principal symbols of the projections $\widehat{P}_+$ and $\widehat{P}_-$ (the Calderón bundles), i.e., the principal symbols of the spaces $\widehat{L}_+$ and $\widehat{L}_-$.

Proof. Theorem 5.4 follows from formula (5.16) applied to problems (5.21) and (5.24) with regard to Theorem 5.1. $\square$

5.2 Agranovich–Dynin Theorem

This theorem, as well as the "dual" Agranovich theorem considered in the next section, expresses the superposition principle for the relative index as applied to boundary value problems. Both theorems were obtained at very early stages of the development of index theory of boundary value problems. The history of the topic is described in [1], where one can also find references to the original papers.

Theorem 5.5. *Let $\widehat{D}$ be an elliptic differential operator on a compact C^∞ manifold X with boundary $\partial X = Y$, and let $\widehat{B}_1$ and $\widehat{B}_2$ be two operators each of which specifies elliptic (in the sense of Shapiro–Lopatinskii) boundary conditions for the operator $\widehat{D}$. Then the relative index of the elliptic boundary value problems $(\widehat{D}, \widehat{B}_1)$ and $(\widehat{D}, \widehat{B}_2)$ is equal to*

$$\operatorname{ind}(\widehat{D}, \widehat{B}_1) - \operatorname{ind}(\widehat{D}, \widehat{B}_2) = \operatorname{ind} \widehat{(B_1 \circ B_2^{-1})} \tag{5.28}$$

where $\widehat{(B_1 \circ B_2^{-1})}$ is an elliptic pseudodifferential operator on Y with principal symbol $(B_1 \circ B_2^{-1})$; here B_1 and B_2 are treated as the restrictions of the principal symbols of $\widehat{B}_1$ and $\widehat{B}_2$ to the subbundle L_+, which is the principal symbol of the subspace $\widehat{L}_+$.

Proof. We shall derive this well-known theorem from the superposition principle for the relative index. We equip Sobolev spaces on X with the structure of collar spaces using a function $\chi : X \to [-1, 1]$ equal to -1 in a neighborhood of Y, equal to 1 outside the collar neighborhood U of Y, and increasing from -1 to 1 in U. In various function spaces on Y, we also introduce the structure of collar spaces by setting

$$\varphi g \stackrel{\text{def}}{=} \varphi(-1)g \tag{5.29}$$

for any elements g of such spaces and any $\varphi \in C^\infty([-1,1])$. Then elliptic boundary value problems generate c-Fredholm operators in collar spaces (this follows from the structure of parametrices of boundary value problems; e.g., see [36]). Without loss of generality, we can assume that the coefficients of $\widehat{D}$ are independent of the collar variable t in U. Consider the diagram of modifications shown in Fig. 5.1.

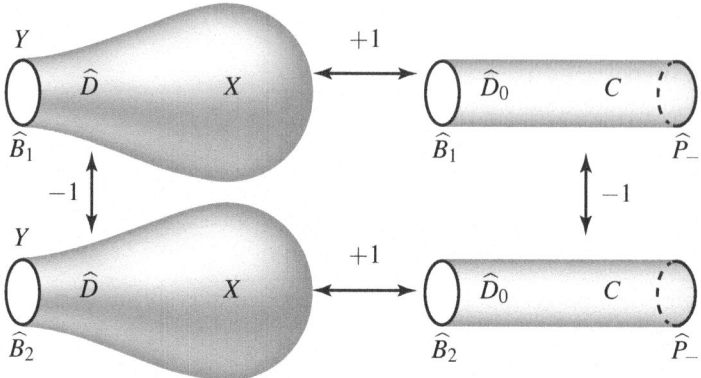

Figure 5.1: Modifications for boundary value problems

Here $\widehat{D}_0$ in the right column of the diagram is the operator on the cylinder naturally obtained from $\widehat{D}$ by freezing the coefficients on the boundary.

By the superposition principle for the relative index, we have

$$\mathrm{ind}(\widehat{D},\widehat{B}_1) - \mathrm{ind}(\widehat{D},\widehat{B}_2) = \mathrm{ind}(\widehat{D}_0,\widehat{B}_1,\widehat{P}_-) - \mathrm{ind}(\widehat{D}_0,\widehat{B}_2,\widehat{P}_-). \tag{5.30}$$

The indices on the right-hand side can be computed by formula (5.16) with regard to the fact that the index of the problem $(\widehat{D}_0,\widehat{P}_+,\widehat{P}_-)$ is zero. We have

$$\mathrm{ind}(\widehat{D}_0,\widehat{B}_1,\widehat{P}_-) = \mathrm{ind}(\widehat{B}_1 : \widehat{L}_+ \to \mathscr{L}_1),$$
$$\mathrm{ind}(\widehat{D}_0,\widehat{B}_2,\widehat{P}_-) = \mathrm{ind}(\widehat{B}_2 : \widehat{L}_+ \to \mathscr{L}_2),$$

where $\mathscr{L}_1$ and $\mathscr{L}_2$ are the Sobolev spaces on Y in which the operators $\widehat{B}_1$ and $\widehat{B}_2$ act. Then

$$\mathrm{ind}(\widehat{D}_0,\widehat{B}_1,\widehat{P}_-) - \mathrm{ind}(\widehat{D}_0,\widehat{B}_2,\widehat{P}_-) = \mathrm{ind}(\widehat{B}_1\widehat{B}_2^{[-1]} : \mathscr{L}_2 \to \mathscr{L}_1) = \mathrm{ind}(\widehat{B_1 B_2^{-1}}), \tag{5.31}$$

as desired. (By $\widehat{B}_2^{[-1]}$ we denote the almost inverse of $\widehat{B}_2$.) □

5.3 Agranovich Theorem

The Agranovich theorem deals in a sense with the opposite case.

Theorem 5.6. *Let $\widehat{D}_1$ and $\widehat{D}_2$ be two elliptic differential operators on a compact C^∞ manifold X with boundary $\partial X = Y$ coinciding in a collar neighborhood of the boundary, and let $\widehat{B}$ be a boundary operator satisfying the Shapiro–Lopatinskii conditions with respect to $\widehat{D}_1$ (and hence with respect to $\widehat{D}_2$). Then the relative index of the problems $(\widehat{D}_1, \widehat{B})$ and $(\widehat{D}_2, \widehat{B})$ is equal to*

$$\operatorname{ind}(\widehat{D}_1, \widehat{B}) - \operatorname{ind}(\widehat{D}_2, \widehat{B}) = \operatorname{ind}(\widehat{D_1 D_2^{-1}}), \tag{5.32}$$

where $\widehat{D_1 D_2^{-1}}$ is a pseudodifferential operator on X with principal symbol $D_1 D_2^{-1}$ acting as the identity operator on functions supported in a sufficiently small neighborhood of the boundary.

Remark 5.7. The operator $D_1 D_2^{-1}$ obviously requires no boundary conditions.

Proof. The operators $\widehat{D}_1$ and $\widehat{D}_2$ can be extended to the double $2X = X \underset{Y}{\cup} X$ as elliptic operators (see [78]).

Since $\widehat{D}_1$ and $\widehat{D}_2$ coincide near the boundary, we can assume that the extensions coincide on the second copy of X. Let us denote these extensions by $\mathfrak{D}_1$ and $\mathfrak{D}_2$. Now consider the modification diagram shown in Fig. 5.2.

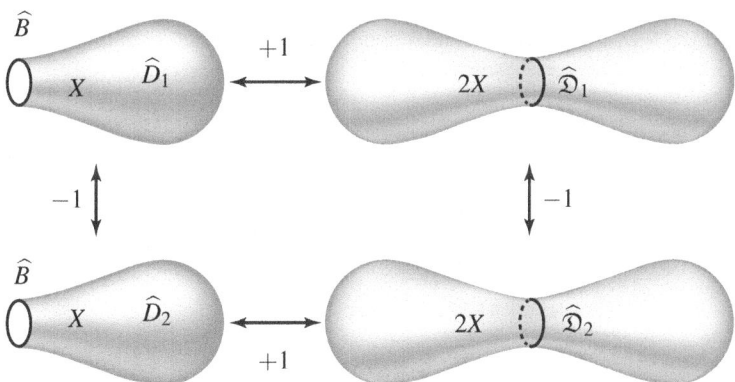

Figure 5.2: Extension to the double

By the superposition principle for the relative index, we obtain

$$\operatorname{ind}(\widehat{D}_1, \widehat{B}) - \operatorname{ind}(\widehat{D}_2, \widehat{B}) = \operatorname{ind}(\mathfrak{D}_1) - \operatorname{ind}(\mathfrak{D}_2) = \operatorname{ind}(\widehat{\mathfrak{D}_1 \mathfrak{D}_2^{-1}}).$$

But it is obvious that

$$\operatorname{ind}(\widehat{\mathfrak{D}_1 \mathfrak{D}_2^{-1}}) = \operatorname{ind}(\widehat{D_1 D_2^{-1}}),$$

since the symbol $\mathfrak{D}_1\mathfrak{D}_2^{-1}$ of the operator $\widehat{\mathfrak{D}_1\mathfrak{D}_2^{-1}}$ is equal to unity on the second copy of X and in a neighborhood of Y, so that this operator can be homotopied to an operator acting as the identity operator on functions supported on the second copy of X or in a neighborhood of Y. □

5.4 Bojarski Theorem and Its Generalizations

In the mid-1970s, Bojarski put forward the following *cutting conjecture* in the framework of a surgery proof of the Atiyah–Singer index theorem, which he was developing at the time. Consider a Dirac operator $\widehat{D}$ on a closed connected manifold M. We cut M by a two-sided hypersurface S into two parts M_+ and M_-, $\partial M_+ = \partial M_- = S$, and equip the resulting Dirac operators on M_+ and M_- with the Atiyah–Patodi–Singer conditions $\widehat{P}_+ u_+ = 0$, $\widehat{P}_- u_- = 0$. Then the index of the Dirac operator on M is equal to the relative index of the Fredholm pair of subspaces

$$(\widehat{L}_+ = \operatorname{Im} \widehat{P}_+, \widehat{L}_- = \operatorname{Im} \widehat{P}_-).$$

Later, this conjecture was proved (the Bojarski theorem); see the book [16] for details. Here we shall prove a theorem on cutting an arbitrary elliptic operator into boundary value problems.

Let M be a closed C^∞ manifold, $\widehat{D}$ an elliptic differential operator on M, and $S \subset M$ a smooth two-sided hypersurface. We cut M along S into two manifolds M_+ and M_- with boundary $\partial M_+ = \partial M_- = S$ and consider general elliptic boundary value problems on M_+ and M_-:

$$\begin{cases} \widehat{D} u_+ = f_+, & \text{on } M_+, \\ \widehat{B}_+ j_S^{m-1} u_+ = g_+ \in \mathscr{L}_+, \end{cases} \quad (5.33)$$

$$\begin{cases} \widehat{D} u_+ = f_-, & \text{on } M_-, \\ \widehat{B}_- j_S^{m-1} u_- = g_- \in \mathscr{L}_-, \end{cases} \quad (5.34)$$

where

$$\widehat{B}_+ : \mathscr{H}_m^s(S) \to \mathscr{L}_+ \quad (5.35)$$

$$\widehat{B}_- : \mathscr{H}_m^s(S) \to \mathscr{L}_-, \quad (5.36)$$

are some operators of boundary conditions such that problems (5.33) and (5.34) are Fredholm and $\mathscr{L}_+$ and $\mathscr{L}_-$ are some Hilbert spaces. The restrictions of $\widehat{B}_+$ to $\widehat{L}_+$ and $\widehat{B}_-$ to $\widehat{L}_-$, where $\widehat{L}_\pm = \operatorname{Im} \widehat{P}_\pm$ and the projections $\widehat{P}_\pm$ correspond to the conormal symbol of the operator $\widehat{D}$, will be denoted by the same letters.

Theorem 5.8.

$$\operatorname{ind} \widehat{D} = \operatorname{ind}(\widehat{D}_+, \widehat{B}_+) + \operatorname{ind}(\widehat{D}_-, \widehat{B}_-) - \operatorname{ind} \begin{pmatrix} \widehat{B}_+ & \widehat{L}_+ & \mathscr{L}_+ \\ \oplus & : \oplus & \longrightarrow & \oplus \\ \widehat{B}_- & \widehat{L}_- & \mathscr{L}_- \end{pmatrix}. \quad (5.37)$$

5.5. Boundary Value Problems with Symmetric Conormal Symbol

Proof. Consider the modification diagram shown in Fig. 5.3.

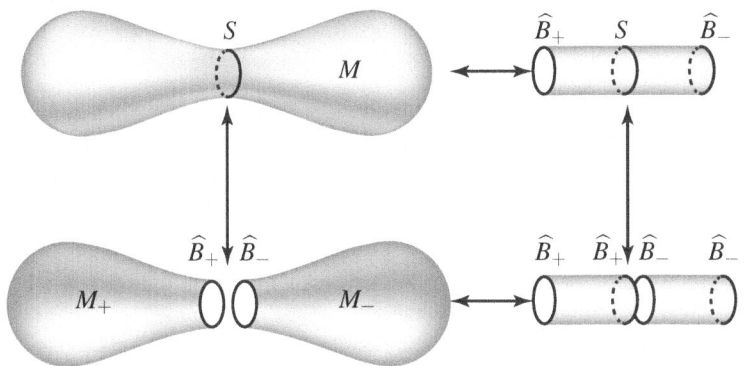

Figure 5.3: Cutting into boundary value problems

In the left column, the main elliptic operator is the operator $\widehat{D}$ on M and its restrictions to M_+ and M_-. In the right column, the main operator is given by the extension to the finite cylinder C of the operator $\widehat{D}$ with coefficients frozen on S. Needless to say, just as before, we assume that the coefficients of $\widehat{D}$ are independent of the collar variable t. By the relative index theorem, we have

$$\operatorname{ind}\widehat{D} - \operatorname{ind}(\widehat{D}_+, \widehat{B}_+) - \operatorname{ind}(\widehat{D}_-, \widehat{B}_-) = \operatorname{ind}\widehat{B}_+ \oplus \widehat{B}_-$$

(the remaining two terms on the right-hand side are zero). The proof of the theorem is complete. $\square$

5.5 Boundary Value Problems with Symmetric Conormal Symbol

On a C^∞ manifold X with boundary $\partial X = Y$, consider the elliptic boundary value problem

$$\begin{cases} \widehat{D}u = f, \\ \widehat{B}j_Y^{m-1}u = g \in \mathscr{L}. \end{cases} \quad (5.38)$$

Suppose that the conormal symbol $\widehat{D}_0(p)$ of the operator

$$\widehat{D} : H^s(X,E) \to H^{s-m}(X,F)$$

satisfies the symmetry condition $\widehat{D}_0(p) = \widehat{D}_0(-p)$. One can also consider the more general case in which the symmetry includes a diffeomorphism $g : Y \to Y$ and associated bundle isomorphisms

$$\mu_E : E|_Y \to g^*E|_Y, \qquad \mu_F : F|_Y \to g^*F|_Y.$$

92   Chapter 5. Applications to Boundary Value Problems

In this case, using the general superposition principle for the relative index, for the index of the boundary value problem (5.38) one can give a simpler formula than the general Atiyah–Bott formula [5].

Indeed, consider the surgery that takes two copies of the operator $\{\widehat{D}, \widehat{B}j_Y^{m-1}\}$ to a new operator $\widehat{\widehat{D}}$ on the closed manifold (double) $2M$ and the operator $\widehat{D}_0 = \widehat{D}_0(-i\partial/\partial t)$ with coefficients independent of t on the cylinder $Y \times [-1/2, 1/2]$ with the boundary conditions

$$\widehat{B} j^{m-1}_{\{-1/2\}} u = g_1, \qquad \widehat{B} j^{m-1}_{\{+1/2\}} u = g_2$$

on the faces.

The index of this model boundary value problem on the cylinder is equal to the index of the spectral problem for $\widehat{D}_0$ plus the index of the operator

$$\begin{pmatrix} \widehat{B} \\ \widehat{B} \end{pmatrix} : \begin{matrix} \widehat{L}_+ \\ \oplus \\ \widehat{L}_- \end{matrix} \to \begin{matrix} \mathscr{L} \\ \oplus \\ \mathscr{L}, \end{matrix}$$

where $\widehat{L}_\pm \subset \bigoplus_{k=0}^{m-1} H^{s-1/2-k}(Y, E|_Y)$ are the Calderón subspaces [23] corresponding to the left and right faces of the cylinder. Without loss of generality (say, adding a constant to $\widehat{D}$, which does not affect the index) we can assume that $\widehat{D}_0(p)$ is invertible for real p. Then the index of the spectral boundary value problem for $\widehat{D}_0$ is zero and

$$\widehat{L}_+ \oplus \widehat{L}_- = \bigoplus_{k=0}^{m-1} H^{s-1/2-k}(Y, E|_Y)$$

by the symmetry condition. Finally, we obtain the following theorem.

Theorem 5.9. *One has the index formula*

$$\mathrm{ind}(\widehat{D}, \widehat{B} \circ j_Y^{m-1}) = \frac{1}{2}\left\{\mathrm{ind}(\widehat{\widehat{D}}) + \mathrm{ind}\left(\widehat{B} \oplus \widehat{B} : \bigoplus_{k=0}^{m-1} H^{s-\frac{1}{2}-k}(Y, E|_Y) \longrightarrow \mathscr{L} \oplus \mathscr{L}\right)\right\}.$$

If the boundary value problem (5.38) is classical, then both terms in this index formula are the indices of elliptic operators on closed manifolds.

Chapter 6

Spectral Flow for Families of Dirac Type Operators with Classical Boundary Conditions

6.1 Statement of the Problem

In this chapter, we discuss the results obtained in [38] with the use of the relative index superposition principle for the spectral flow of families of self-adjoint Dirac type operators with classical elliptic boundary conditions on a compact Riemannian manifold with boundary. In the two-dimensional case, such operators arise, for example, when describing electron states in graphene [39] or in topological insulators [33, 68], and their spectral flow has an important physical meaning, being related to the creation of electron–hole (or, more generally, particle–antiparticle) pairs.

Recall [16, Definition 3.1] that a first-order linear differential operator

$$D\colon C^\infty(M,E) \longrightarrow C^\infty(M,E)$$

on the space of sections of a complex vector bundle E over a Riemannian manifold M is called a *Dirac type operator* if its principal symbol

$$\sigma_D\colon \pi^*E \longrightarrow \pi^*E,$$

where π^*E is the pullback of E to the cotangent bundle $T_0^*M = T^*M \setminus \{0\}$ of M without the zero section $\{0\}$ via the natural projection $\pi\colon T^*M \setminus \{0\} \longrightarrow M$, satisfies the condition

$$(\sigma_D(x,\xi))^2 = \sum_{j,k} g^{jk}(x)\xi_j\xi_k I,$$

where I is the identity operator in the fiber E_x and the $g^{jk}(x)$ are the contravariant components of the Riemannian metric tensor.

Assume that M is a compact manifold with boundary ∂M, E is a Hermitian vector bundle (i.e., there is a Hermitian inner product $\langle \cdot, \cdot \rangle$ defined on each fiber E_x and smoothly depending on x) and further the operator D is *symmetric*, i.e., satisfies

$$(u, Dv) = (Du, v) \qquad \text{for any } u, v \in C_0^\infty(M \setminus \partial M, E),$$

where

$$(u, v) = \int_M \langle u(x), v(x) \rangle \, d\operatorname{vol}(x)$$

is the inner product on the space $L^2(M, E)$ of square integrable sections of E. (Here d vol is the Riemannian volume element.) Under certain conditions, one can equip D with homogeneous boundary conditions to obtain an unbounded self-adjoint Fredholm operator on $L^2(M, E)$. Namely, assume that E is even-dimensional and the restriction $E_{\partial M}$ of E to the boundary of M contains a subbundle $L \subset E_{\partial M}$ of dimension $\dim L = \frac{1}{2} \dim E$ such that

$$E_{\partial M} = L \oplus \sigma_D(\mathbf{n}) L \qquad \text{(orthogonal direct sum of bundles)},$$

where $\mathbf{n} = \mathbf{n}(x)$, $x \in \partial M$, is the unit inward conormal to the boundary ∂M and we write $\sigma_D(\mathbf{n}) = \sigma_D(x, \mathbf{n}(x))$, $x \in \partial M$, for brevity. Let

$$\pi_L \colon E_{\partial M} \longrightarrow E_{\partial M}/L \simeq \sigma_D(\mathbf{n}) L$$

be the natural projection onto the quotient bundle. We claim that *the boundary condition $\pi_L(u|_{\partial M}) = 0$ makes D a self-adjoint Fredholm operator*. More precisely, the following assertion holds.

Proposition 6.1 (see [16, Chaps. 18 and 19] and [20]). *Under the above-mentioned conditions, the operator D with domain consisting of smooth sections $u \in C^\infty(M, E)$ satisfying the condition $\pi_L(u|_{\partial M}) = 0$ is essentially self-adjoint, and its closure is an unbounded Fredholm operator with discrete spectrum on $L^2(M, E)$.*

This operator will be denoted by (D, L) in what follows, and, by slight abuse of language, we refer to L as a *compatible boundary condition for D*. See the cited papers for the precise description of the domain of (D, L).

Example 6.2. The operator

$$A = \begin{pmatrix} 0 & -i\frac{\partial}{\partial x} - \frac{\partial}{\partial y} \\ -i\frac{\partial}{\partial x} + \frac{\partial}{\partial y} & 0 \end{pmatrix} \tag{6.1}$$

acting on sections $u = {}^t(u_1, u_2)$ of the trivial bundle $\mathbb{C}^2$ over the space $\mathbb{R}^2$ with coordinates (x, y) equipped with the standard Euclidean metric $g = dx^2 + dy^2$ is a Dirac type operator, because

$$(\sigma_A(x, \xi))^2 \equiv \begin{pmatrix} 0 & \xi_x - i\xi_y \\ \xi_x + i\xi_y & 0 \end{pmatrix}^2 = (\xi_x^2 + \xi_y^2)\begin{pmatrix} 1 & 0 \\ 0 & 1 \end{pmatrix}.$$

6.1. Statement of the Problem

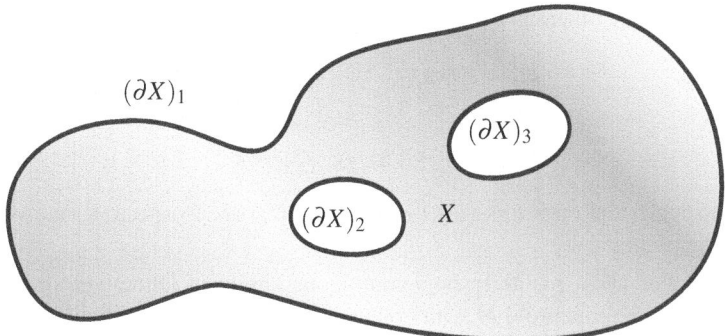

Figure 6.1: Domain with two holes

This operator is also obviously formally self-adjoint (for the standard Hermitian product on $\mathbb{C}^2$). Consider this operator in a bounded domain $X \subset \mathbb{R}^2$ with smooth boundary ∂X, possibly consisting of several connected components,

$$\partial X = (\partial X)_1 \cup (\partial X)_2 \cup \cdots \cup (\partial X)_m$$

(see Fig. 6.1).

Physically, one can interpret X as a graphene sheet (possibly with holes), and the operator A approximates the tight-binding electronic Hamiltonian near the conical self-crossing point K in the first Brillouin zone of the reciprocal lattice [39, Chap. 1]. The boundary conditions of the form described in Proposition 6.1 can be written as

$$(n_y - in_x)u_1|_{\partial X} = Bu_2|_{\partial X}, \tag{6.2}$$

where n_x and n_y are the components of the inward conormal $\mathbf{n}$ to the boundary ∂X and B is a nonvanishing real-valued function on the boundary. Thus, B is of constant sign on each of the boundary components. These boundary conditions were introduced in [12] and accordingly are known as the *Berry–Mondragon boundary conditions*.

Now assume that we are given a *family* (D_t, L_t), $t \in [0, 1]$, of self-adjoint Fredholm Dirac type operators of the kind described in Proposition 6.1, where the coefficients of the operator D_t and the subbundle $L \subset E|_{\partial M}$ continuously depend on t. Then the operator family (D_t, L_t) is continuous in the sense of strong resolvent convergence [15, Theorem 7.16], and the *spectral flow* $\operatorname{sf}\{(D_t, L_t)\}$ of this family is well defined (e.g., see [14]). Recall that, informally speaking, the spectral flow of a family $\{C_t\}$, $t \in [0, 1]$, of self-adjoint operators with discrete spectrum is the "net number of eigenvalues that cross zero in the positive direction" as t varies from 0 to 1. (See the exact definition later on in this chapter.) If we consider only families for which the operators C_0 and C_1 are *isospectral* (i.e., have the same eigenvalues, counting multiplicities), then the spectral flow proves to be a homotopy invariant, and its computation is of interest. Returning to our Dirac type operators, assume that there is a bundle automorphism $U: E \longrightarrow E$ such that

$$D_1 = UD_0U^{-1}, \qquad L_1 = U(L_0). \tag{6.3}$$

Then

$$(D_1, L_1) = U(D_0, L_0)U^{-1};$$

hence in particular the operators (D_0, U_0) and (D_1, U_1) are isospectral, and we arrive at the following problem.

Problem 6.3. Let $\{D_t\}$, $t \in [0,1]$, be a continuous family of Dirac type operators on a compact Riemannian manifold M with boundary, and let $\{L_t\}$ be a continuous family of compatible boundary conditions for D_t. Next, let U be a bundle automorphism such that relations (6.3) are satisfied. Find the spectral flow $\text{sf}\{(D_t, L_t)\}$.

Example 6.4. Continuing Example 6.2, let $U(x,y)$ be a smooth complex-valued function on X with $|U(x,y)| \equiv 1$, so that locally one has

$$U(x,y) = e^{iS(x,y)}, \qquad \text{where } S(x,y) \text{ is smooth and real-valued.}$$

Then, denoting the operator of multiplication by U by the same letter, we obtain

$$UAU^{-1} = \begin{pmatrix} 0 & -i\frac{\partial}{\partial x} - \frac{\partial}{\partial y} + Q(x,y) \\ -i\frac{\partial}{\partial x} + \frac{\partial}{\partial y} + \overline{Q}(x,y) & 0 \end{pmatrix},$$

where

$$Q(x,y) = -\frac{\partial S}{\partial x}(x,y) + i\frac{\partial S}{\partial y}(x,y).$$

The subbundle L determined by the boundary condition (6.2) is invariant under U (because U is a scalar transformation), and we can consider the family $\{(A_t, L)\}$ of self-adjoint Fredholm operators, where

$$A_t = tUAU^{-1} + (1-t)A = \begin{pmatrix} 0 & -i\frac{\partial}{\partial x} - \frac{\partial}{\partial y} + tQ(x,y) \\ -i\frac{\partial}{\partial x} + \frac{\partial}{\partial y} + t\overline{Q}(x,y) & 0 \end{pmatrix}.$$

(Note that $\sigma_{A_t} = \sigma_A$ is independent of t.)

Physically, this transformation of the operator A is a gauge transformation, the gradient of S can be interpreted as the electromagnetic vector potential, and the spectral flow of the family $\{(A_t, L)\}$ is the number of electron–hole pairs created. The winding numbers

$$\text{wind}_j(U) = \frac{1}{2\pi} \oint_{(\partial X)_j} \frac{dU}{U}$$

of the restrictions of U to the boundary components can physically be interpreted as the numbers of magnetic flux quanta through the respective holes in the graphene sheet. They are the only homotopy invariants of U, and hence it is natural to expect that the solution of Problem 6.3 in this particular case expresses $\text{sf}\{(A_t, L)\}$ via these winding numbers.

6.2. Simple Example

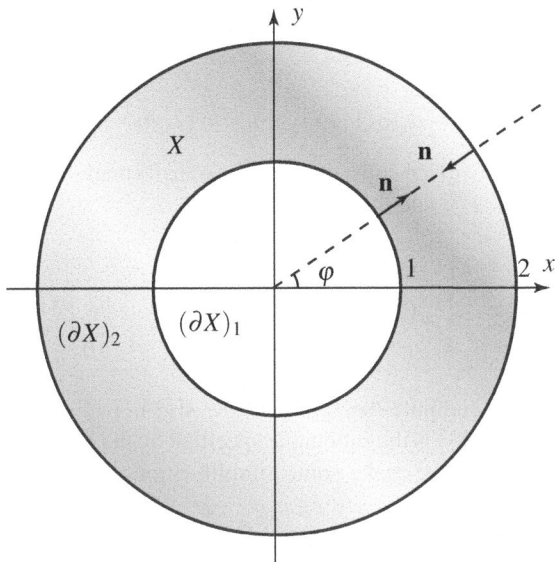

Figure 6.2: Annulus

The outline of the remaining part of this chapter is as follows. First, we find the spectral flow by a straightforward computation in a simple example (Section 6.2). Then we state the main theorem in [38] and briefly explain how it is proved with the use of the relative index superposition principle (Section 6.3). Finally, we show how this theorem applies to compute the spectral flow in our Example 6.4, whereby we once again use the relative index superposition principle to reduce the general case of a domain with m holes to the example considered in Section 6.2.

6.2 Simple Example

Consider the operator (6.1) with boundary conditions of the form (6.2) in the simplest possible domain where the spectral flow can be nontrivial, namely, in the annulus $1 \leq \rho \leq 2$ on the (x,y)-plane, where $\rho = (x^2 + y^2)^{1/2}$ (see Fig. 6.2). We will use the polar coordinates (ρ, φ), in which the operator A acquires the form

$$A = \begin{pmatrix} 0 & -ie^{-i\varphi}\frac{\partial}{\partial \rho} - \frac{1}{\rho}e^{-i\varphi}\frac{\partial}{\partial \varphi} \\ -ie^{i\varphi}\frac{\partial}{\partial \rho} + \frac{1}{\rho}e^{i\varphi}\frac{\partial}{\partial \varphi} & 0 \end{pmatrix}. \qquad (6.4)$$

The inward normal to the boundary has the form

$$\mathbf{n}(\varphi) = (\cos\varphi, \sin\varphi) \quad \text{on } (\partial X)_1, \qquad \mathbf{n}(\varphi) = (-\cos\varphi, -\sin\varphi) \quad \text{on } (\partial X)_2.$$

We take $B = \pm 1$ in the boundary conditions (6.2) (all other nonzero real values of B can be continuously deformed to these without violating self-adjointness and ellipticity and hence without affecting the spectral flow); then the boundary conditions for the operator A acquire the form

$$ie^{i\varphi}u_1(1,\varphi) = B_1 u_2(1,\varphi), \quad -ie^{i\varphi}u_1(2,\varphi) = B_2 u_2(2,\varphi), \tag{6.5}$$

where the $B_j \in \{-1,1\}$ are given constants. For the transformation U we take the operator of multiplication by $U = e^{i\varphi}$, and then

$$A_t = tUAU^{-1} + (1-t)A$$

$$= \begin{pmatrix} 0 & -ie^{-i\varphi}\frac{\partial}{\partial \rho} - \frac{1}{\rho}e^{-i\varphi}\frac{\partial}{\partial \varphi} + \frac{it}{\rho}e^{-i\varphi} \\ -ie^{i\varphi}\frac{\partial}{\partial \rho} + \frac{1}{\rho}e^{i\varphi}\frac{\partial}{\partial \varphi} - \frac{it}{\rho}e^{i\varphi} & 0 \end{pmatrix}. \tag{6.6}$$

Thus, the problem is to compute the spectral flow sf$\{(A_t, L)\}$ of the family $\{(A_t, L)\}$, $t \in [0,1]$, where $L = L(B_1, B_2)$ is the subbundle specified by the boundary conditions (6.5).

For this computation, we make some simplifications. First, let us introduce new unknown functions $v = {}^t(v_1, v_2)$ by setting $u_1 = -ie^{-i\varphi}v_1$, $u_2 = v_2$. Then we obtain the new, unitarily equivalent family of Dirac type operators

$$D_t = \begin{pmatrix} ie^{i\varphi} & 0 \\ 0 & 1 \end{pmatrix} A_t \begin{pmatrix} -ie^{-i\varphi} & 0 \\ 0 & 1 \end{pmatrix} = \begin{pmatrix} 0 & \frac{\partial}{\partial \rho} - \frac{i}{\rho}\frac{\partial}{\partial \varphi} - \frac{t}{\rho} \\ -\frac{\partial}{\partial \rho} - \frac{i}{\rho}\frac{\partial}{\partial \varphi} - \frac{t+1}{\rho} & 0 \end{pmatrix} \tag{6.7}$$

with the boundary conditions

$$v_1(1,\varphi) = B_1 v_2(1,\varphi), \quad -v_1(2,\varphi) = B_2 v_2(2,\varphi). \tag{6.8}$$

The family (6.7) of Dirac type operators is associated with the Riemannian metric and volume form

$$dr^2 = d\rho^2 + \rho^2 d\varphi^2, \quad d\text{vol} = \rho \, d\rho \wedge d\varphi.$$

Let us further simplify the problem by deforming the operator family (6.7) to a family of operators with constant coefficients. For this to work, we have to deform the metric simultaneously in an appropriate way. Specifically, for $\alpha \in [0,1]$ we set

$$dr_\alpha^2 = d\rho^2 + \rho^{2\alpha} d\varphi^2, \quad d\text{vol}_\alpha = \rho^\alpha d\rho \wedge d\varphi, \tag{6.9}$$

$$D_{t,\alpha} = \begin{pmatrix} 0 & \frac{\partial}{\partial \rho} - \frac{i}{\rho^\alpha}\frac{\partial}{\partial \varphi} - \frac{t}{\rho^\alpha} \\ -\frac{\partial}{\partial \rho} - \frac{i}{\rho^\alpha}\frac{\partial}{\partial \varphi} - \frac{t}{\rho^\alpha} - \frac{\alpha}{\rho} & 0 \end{pmatrix}. \tag{6.10}$$

For $\alpha = 1$, this is just the operator family (6.7). For each $\alpha \in [0,1]$, the following is true.

1. The operator $D_{t,\alpha}$ is symmetric in $L^2(X, \mathbb{C}^2, d\text{vol}_\alpha)$.

2. One has $(\sigma_{D_{t,\alpha}})^2 = (p_\rho^2 + \rho^{-2\alpha} p_\varphi^2)I$, and so $D_{t,\alpha}$ is a Dirac type operator for the metric (6.9).

3. On the boundary ∂X, one has

$$\sigma_{D_{t,\alpha}}(\mathbf{n}) = \pm \begin{pmatrix} 0 & i \\ -i & 0 \end{pmatrix}$$

6.2. Simple Example

(the upper sign corresponds to $(\partial X)_1$), and hence $\sigma_{D_{t,\alpha}}(\mathbf{n})$ takes the subspace of vectors of the form ${}^t(Bw,w)$ with fixed real B to its orthogonal complement; this shows that the boundary conditions (6.8) are compatible with $D_{t,\alpha}$ and hence the family $(D_{t,\alpha},L)$ of the operators (6.10) with these boundary conditions consists of self-adjoint Fredholm operators in $L^2(X,\mathbb{C}^2,d\operatorname{vol}_\alpha)$.

4. For each α, one has $D_{1,\alpha} = UD_{0,\alpha}U^{-1}$, and so the starting (for $t=0$) and final (for $t=1$) operators in the family $(D_{t,\alpha},L)$ are isospectral.

Being homotopy invariant, the spectral flow of the family $(D_{t,\alpha},L)$ is independent of α, and it suffices to compute the spectral flow for the family $(D_{t,0},L)$, where

$$D_{t,0} = \begin{pmatrix} 0 & \frac{\partial}{\partial \rho} - i\frac{\partial}{\partial \varphi} - t \\ -\frac{\partial}{\partial \rho} - i\frac{\partial}{\partial \varphi} - t & 0 \end{pmatrix}. \tag{6.11}$$

We arrive at the following eigenvalue problem, where $\lambda \in \mathbb{R}$:

$$D_{t,0}v = \lambda v, \qquad (v_1 - B_1 v_2)|_{\rho=1} = 0, \quad (v_2 + B_2 v_2)|_{\rho=2} = 0. \tag{6.12}$$

Seeking the solution by the Fourier method in the form

$$v(\rho,\varphi) = w(\rho)e^{im\varphi}, \qquad m \in \mathbb{Z},$$

we obtain the following system of ordinary differential equations with boundary conditions for $w = {}^t(w_1,w_2)$:

$$\frac{\partial w}{\partial \rho} = \begin{pmatrix} m-t & -\lambda \\ \lambda & t-m \end{pmatrix} w, \qquad w(1) = \begin{pmatrix} B_1 \\ 1 \end{pmatrix}, \quad w(2) = c\begin{pmatrix} -B_2 \\ 1 \end{pmatrix}, \tag{6.13}$$

the coefficient $c \in \mathbb{C}$ being so far arbitrary. The matrix of the system in (6.13) has the property

$$\begin{pmatrix} m-t & -\lambda \\ \lambda & t-m \end{pmatrix}^2 = ((m-t)^2 - \lambda^2)\begin{pmatrix} 1 & 0 \\ 0 & 1 \end{pmatrix},$$

and hence we obtain

$$w(2) = e^{\begin{pmatrix} m-t & -\lambda \\ \lambda & t-m \end{pmatrix}} w(1) = \cosh\mu\, w(1) + \frac{\sinh\mu}{\mu}\begin{pmatrix} m-t & -\lambda \\ \lambda & t-m \end{pmatrix} w(1),$$

where $\mu = \sqrt{(m-t)^2 - \lambda^2}$, or

$$w(2) = \begin{cases} \cosh\mu \begin{pmatrix} 1 \\ 1 \end{pmatrix} + \dfrac{\sinh\mu}{\mu}(t-m+\lambda)\begin{pmatrix} -1 \\ 1 \end{pmatrix} & \text{if } B_1 = 1, \\ \cosh\mu \begin{pmatrix} -1 \\ 1 \end{pmatrix} + \dfrac{\sinh\mu}{\mu}(t-m-\lambda)\begin{pmatrix} 1 \\ 1 \end{pmatrix} & \text{if } B_1 = -1. \end{cases} \tag{6.14}$$

Since we wish to compute the spectral flow, we are interested only in eigenvalues λ sufficiently small in modulus. Then the factors $\cosh\mu$ and $\sinh\mu/\mu$ in (6.14) are strictly positive regardless of whether $(m-t)^2 - \lambda^2$ is positive or negative. For λ to be an eigenvalue, we need the last relation in (6.13) to be satisfied. Since $\cosh\mu \neq 0$, this is only possible if $B_2 = -B_1$. Thus, there is no spectral flow at all if $B_2 = B_1$. Now if $B_2 = -B_1$, then the last relation in (6.13) implies that the second term in the sum (6.14) is zero. Since $\sinh\mu/\mu \neq 0$, we obtain

$$\lambda = \lambda_m(t) = \begin{cases} m-t & \text{if } B_1 = -B_2 = 1, \\ t-m & \text{if } B_1 = -B_2 = -1. \end{cases} \quad (6.15)$$

By adding a small number ε to the operators $D_{t,0}$ so as to ensure that $D_{0,0}$ and $D_{1,0}$ are invertible, we can readily find the spectral flow. Indeed, there is only one value of m for which the eigenvalue crosses zero, and we arrive at the following assertion.

Proposition 6.5. *The spectral flow of the operator family* (6.6) *with the boundary conditions* (6.5) *is given by the formula*

$$\mathrm{sf}\{(A_t, L(B_1, B_2))\} = \begin{cases} 0 & \text{if } B_1 = B_2, \\ -1 & \text{if } B_1 = -B_2 = 1, \\ 1 & \text{if } B_1 = -B_2 = -1. \end{cases} \quad (6.16)$$

One can readily express this in terms of the winding numbers of the restrictions of $U = e^{i\varphi}$ to the boundary components. Namely, taking into account the positive sense of the boundary components, we have

$$\mathrm{wind}_j(U) = (-1)^j, \qquad j = 1, 2,$$

and so

$$\mathrm{sf}\{(A_t, L)\} = \sum_{j:\, B_j > 0} \mathrm{wind}_j(U). \quad (6.17)$$

We will see in Section 6.4 that the very same formula (6.17) holds in the general situation of Example 6.4. The proof is based on the representation of the spectral flow as the index of some elliptic operator on $X \times S^1$ (which is explained in the next section) and then on an application of the relative index superposition principle, which permits one to cut a domain with m holes into m domains homotopic to the annulus, for which we have just carried out the computations.

6.3 Formula for the Spectral Flow

Let us present the solution [38] of Problem 6.3. We omit all technical details, however important (they can be retrieved from the original paper), and focus ourselves on how the relative index superposition principle helps one to compute the spectral flow.

6.3. Formula for the Spectral Flow

First, let us note that there is a natural candidate for the spectral flow formula. Namely, Atiyah, Patodi, and Singer [7, p. 95] proved that if $D_t, t \in S^1$, is a periodic family of first-order self-adjoint elliptic operators on a closed compact Riemannian manifold M, then

$$\mathrm{sf}\{D_t\} = \mathrm{ind}\left(\frac{\partial}{\partial t} + D_t\right),$$

where $\partial/\partial t + D_t$ is the elliptic operator on the product $M \times S_1$ obtained by identifying the endpoints of the interval $[0,1]$ to form the circle S^1. (The generalization to the case of a family $D_t, t \in [0,1]$, in which D_0 and D_1 do not coincide but are only conjugate, $D_1 = UD_0U^{-1}$ for some bundle automorphism U, is essentially trivial.)

It turns out that the very same formula holds in our case, where M is a manifold with boundary. Assume that conditions (6.3) are satisfied. Then the operator $\partial/\partial t + D_t$ naturally acts on the sections of the bundle $\mathscr{E}$ over $M \times S_1$ obtained from E by lifting from M to $M \times [0,1]$ and then by clutching with the use of the automorphism U when pasting together the endpoints of $[0,1]$. It is easily seen that this operator is elliptic (i.e., its symbol is a bundle automorphism on $T^*(M \times S_1) \setminus \{0\}$).

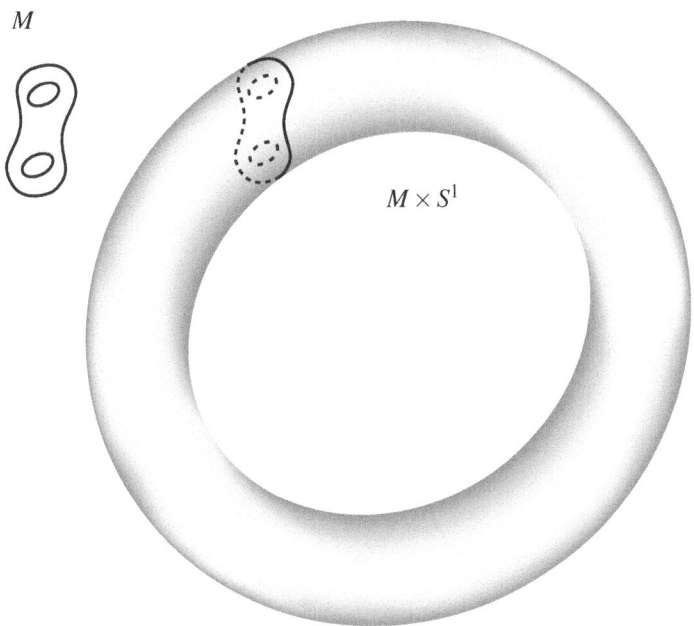

Figure 6.3: The manifold M and the product $M \times S^1$

Next, $M \times S_1$ is a manifold with boundary $\partial M \times S_1$ (see Fig. 6.3), and so we should pose some boundary conditions for the operator $\partial/\partial t + D_t$. This is however easy: the family of subbundles $L_t \subset E_{\partial M} \simeq \mathscr{E}_{\partial M \times \{t\}}$ specifies a well-defined subbundle $\mathscr{L} \subset \mathscr{E}_{\partial M \times S^1}$.

A straightforward computation shows that this subbundle defines a boundary condition of Shapiro–Lopatinskii type (e.g., see [36]), which makes the operator $\partial/\partial t + D_t$ Fredholm.

Now we are in a position to state the main result of this chapter.

Theorem 6.6 ([38, Theorem 2]).

$$\mathrm{sf}\{D_t, L_t\} = \mathrm{ind}\left(\frac{\partial}{\partial t} + D_t, \mathscr{L}\right).$$

The proof is divided into two major steps, and the relative index superposition theorem plays a crucial role at each of these steps.

Step 1. The index of the operator on the product $M \times S^1$ is shown to be equal to the index of the (essentially the same) operator on the infinite cylinder $M \times \mathbb{R}$. The passage from one operator to the other is illustrated in Fig. 6.4.

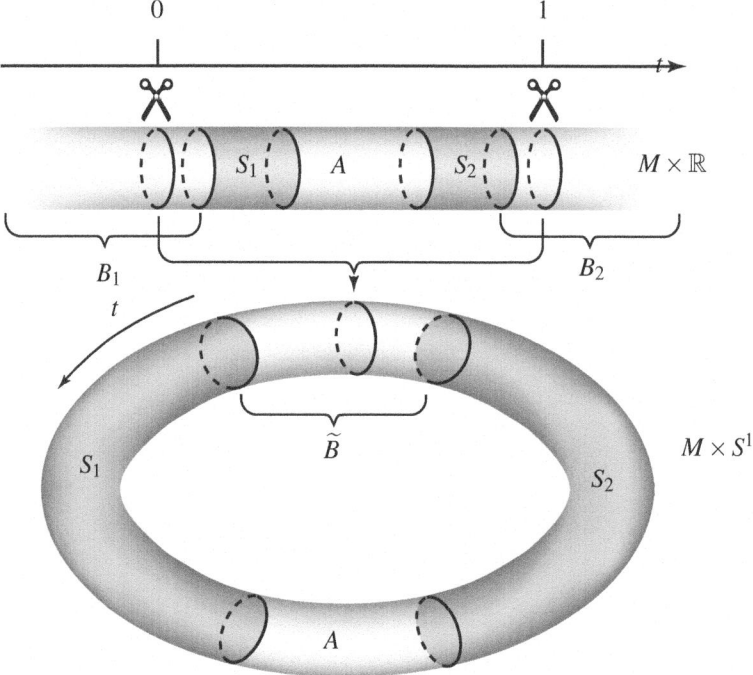

Figure 6.4: From the "torus" to the infinite cylinder

Here some explanations are in order. For simplicity, we assume that 0 is not in the spectrum of (D_0, L_0) (and hence of (D_1, L_1); this can always be achieved by adding a small real constant to the operator family $\{D_t\}$ (which does not affect the spectral flow). Next, by using a homotopy, we can ensure that the operator D_t and the subbundle L_t depend on t only near the midpoint of the interval $[0,1]$ (say, for $t \in [\frac{1}{4}, \frac{3}{4}]$). Then we extend the

6.3. Formula for the Spectral Flow

operator D_t and the subbundle L_t to all $t \in \mathbb{R}$ assuming that they are constant (independent of t) on $(-\infty, \frac{1}{4}]$ as well as on $[\frac{3}{4}, \infty)$. Hence we obtain a well-defined operator

$$\mathscr{D} = \left(\frac{\partial}{\partial t} + D_t, L_t\right) \quad \text{on } L^2(M \times \mathbb{R}, E). \tag{6.18}$$

By construction, the coefficients of this operator (and the boundary conditions) depend on t only in A and are independent of t on $B_1 \cup S_1$ as well as on $B_2 \cup S_2$. Moreover, this operator is Fredholm. (This is where we need 0 not to be in the spectrum of (D_0, L_0) and (D_1, L_1).) Then we cut away the infinite ends of the cylinder from $t = 0$ to $t = -\infty$ and from $t = 1$ to $t = \infty$ and paste together the end faces of the resulting finite cylinder to obtain the product $M \times S^1$. If we use the clutching homomorphism U to paste together L_0 and L_1, thus obtaining the subbundle $\mathscr{L} \subset \mathscr{E}$, then we arrive at the operator

$$\mathscr{D}_+ = \left(\frac{\partial}{\partial t} + D_t, \mathscr{L}\right) \quad \text{on } L^2(M \times S^1, \mathscr{E}),$$

whose index occurs on the right-hand side in Theorem 6.6.

But why are the indices of the two operators, one on the cylinder and the other on the "torus," the same? It is here that the relative index superposition theorem comes into play. We make the spaces $M \times \mathbb{R}$ and $M \times S^1$ into collar spaces as follows (see Fig. 6.4). We split the cylinder into the domains A, $B = B_1 \cup B_2$ and $S = S_1 \cup S_2$ and the "torus" into the domains $\widetilde{A}$, $\widetilde{B}$ and $S = S_1 \cup S_2$. Next, we map $M \times \mathbb{R}$ and $M \times S^1$ into $[0, 1]$ in such a way that

(i) The mapping, say, f, is differentiable.

(ii) f depends on t alone.

(iii) $f^{-1}(\{-1\}) = A$, $f^{-1}(\{1\}) = B$ (or $\widetilde{B}$), and $f^{-1}((-1, 1)) = S$.

This map defines an action of $C^\infty([-1, 1])$ on $L^2(M \times \mathbb{R}, E)$ and on $L^2(M \times S^1, \mathscr{E})$, making these spaces collar spaces in the sense of Chapter 1.

We see that the passage from the operator on the cylinder to the operator on the "torus" is a modification at 1 in the sense of Theorem 0.10. By that theorem, the index increment resulting from such a modification is independent of what the operators look like at -1, i.e., on A. Let us modify our operators on A so as to obtain new operators for which the index increment is clearly zero. We obtain such operators as follows: on the cylinder $M \times \mathbb{R}$, take

$$\mathscr{D}_- = \begin{cases} (\frac{\partial}{\partial t} + D_0, L_0), & t \leq \frac{1}{2}, \\ (\frac{\partial}{\partial t} + D_1, L_1), & t \geq \frac{1}{2}. \end{cases} \tag{6.19}$$

Here we use the clutching automorphism U to paste together the bundles where the operators act at $t = \frac{1}{2}$; in view of (6.3), the resulting operator, as well as the boundary conditions, continuously (and even smoothly) depends on t. Note that $\mathscr{D}_-$ differs from $\mathscr{D}$ only on A. Now we make the same surgery on $\mathscr{D}_-$ as was previously done on $\mathscr{D}$, thus obtaining an operator $\mathscr{D}_\pm$ on $M \times S_1$. Thus, we have the surgery diagram

$$\begin{array}{ccc} \mathscr{D} & \xrightarrow{1} & \mathscr{D}_+ \\ {\scriptstyle -1}\updownarrow & & \updownarrow{\scriptstyle -1} \\ \mathscr{D}_- & \xrightarrow[1]{} & \mathscr{D}_\pm \end{array}$$

and
$$\operatorname{ind}\mathscr{D} - \operatorname{ind}\mathscr{D}_+ = \operatorname{ind}\mathscr{D}_- - \operatorname{ind}\mathscr{D}_\pm$$

by Theorem 0.10. But
$$\operatorname{ind}\mathscr{D}_- = \operatorname{ind}\mathscr{D}_\pm = 0.$$

To see why, note that (6.19) is just different notation for the operator

$$\left(\frac{\partial}{\partial t}+D_0, L_0\right) \qquad \text{on } L^2(M \times \mathbb{R}, E_0). \tag{6.20}$$

Indeed, the only difference is that, for $t \geq \frac{1}{2}$, different coordinates are used in (6.19) in the fibers of the bundle $\mathscr{E}$. Likewise, the operator $\mathscr{D}_\pm$ is just a different representation of the operator (see Fig. 6.5)

$$\left(\frac{\partial}{\partial t}+D_0, L_0\right) \qquad \text{on } L^2(M \times S^1, E_0). \tag{6.21}$$

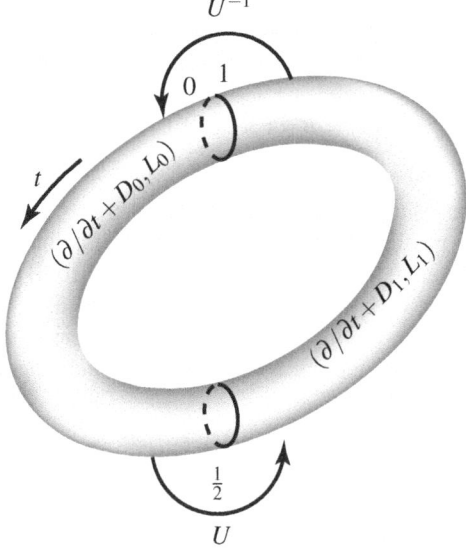

Figure 6.5: Trivial operator on the "torus"

6.3. Formula for the Spectral Flow

The index of each of the operators (6.20) and (6.21) is zero. Indeed, since (D_0, L_0) is self-adjoint and its resolvent set contains a neighborhood of zero, it is an easy exercise to show that both operators are invertible.

Thus, it suffices to prove that the desired spectral flow coincides with the index of the operator (6.18).

Step 2. Let us prove that

$$\mathrm{sf}\{D_t, L_t\} = \mathrm{ind}\left\{\left(\frac{\partial}{\partial t} + D_t, L_t\right) \text{ on } L^2(M \times \mathbb{R}, E)\right\}. \tag{6.22}$$

The definition of spectral flow in the form used in [38] (and originating from [14, 48]) says that (assuming that (D_0, L_0) and (D_1, L_1) are invertible)

$$\mathrm{sf}\{D_t, L_t\} = \sum_{j=1}^{n} m_j \, \mathrm{sign}(\gamma_j - \gamma_{j+1}), \tag{6.23}$$

where $0 = \gamma_1, \gamma_2, \ldots, \gamma_n, \gamma_{n+1} = 0$ is a finite sequence of real constants such that, for some partition $0 = t_0 < t_1 < \cdots < t_{n+1} = 1$ of the interval $[0, 1]$, the number γ_j is contained in the resolvent set of (D_t, L_t) for each $t \in [t_{j-1}, t_j]$ and m_j is the total multiplicity of the eigenvalues of (D_{t_j}, L_{t_j}) lying between γ_j and γ_{j+1}.

Using this definition, one can prove (6.22) by some sort of induction on n. Namely, define the operator

$$\mathscr{D}_j = \begin{cases} (\frac{\partial}{\partial t} + D_0, L_0), & t \leq 0, \\ (\frac{\partial}{\partial t} + D_t, L_t), & t \in [0, t_j], \\ (\frac{\partial}{\partial t} + D_{t_j}, L_{t_j}), & t \geq t_j. \end{cases} \tag{6.24}$$

The operator $\mathscr{D}_0$ is an invertible operator in $L^2(M \times \mathbb{R}, E)$, and $\mathscr{D}_{n+1}$ is the desired operator whose index we need to compute. We use the following trick. Let us introduce the weighted L^2-space $L^2_\gamma(M \times \mathbb{R}, E)$ with norm

$$\|u\|_\gamma = \left\{ \int_{-\infty}^{0} \|u(t)\|^2_{L^2(M,E)} \, dt + \int_{0}^{\infty} \|u(t)\|^2_{L^2(M,E)} e^{2\gamma t} \, dt \right\}^{1/2}.$$

This is the usual L^2-norm on the negative part $t < 0$ of the cylinder and the L^2-norm with weight $e^{\gamma t}$ on the positive part. Since the numbers γ_j and γ_{j+1} belong to the resolvent set of the operator (D_{t_j}, L_{t_j}), it follows that the operator $\mathscr{D}_j$ is Fredholm both in $L^2_{\gamma_j}(M \times \mathbb{R}, E)$ and in $L^2_{\gamma_{j+1}}(M \times \mathbb{R}, E)$. We denote these operators by $\mathscr{D}_j^\gamma$, $\gamma = \gamma_j$ or γ_{j+1}, respectively. Now the induction goes on the basis of two facts:

(i) $\mathrm{ind}\,\mathscr{D}_j^{\gamma_j} = \mathrm{ind}\,\mathscr{D}_{j-1}^{\gamma_j}$. This is easy, because these two operators are homotopic in the class of Fredholm operators owing to the condition that γ_j is in the resolvent set of (D_t, L_t) for $t \in [t_{j-1}, t_j]$.

(ii) $\mathrm{ind}\,\mathscr{D}_j^{\gamma_{j+1}} - \mathrm{ind}\,\mathscr{D}_j^{\gamma_j} = m_j \, \mathrm{sign}(\gamma_j - \gamma_{j+1})$. This relative index theorem, which is a straightforward analog of relative index theorems for elliptic operators on manifolds with conical singularities (e.g., see [48], where further references can be found), can be proved with the use of the relative index superposition principle as shown below. (For operators on manifolds with conical singularities, this approach was used in [48].)

Let us introduce the weighted space $L^2_{\nu\gamma}(M \times \mathbb{R}, E)$ with norm

$$\|u\|_\gamma = \left\{ \int_{-\infty}^{0} \|u(t)\|^2_{L^2(M,E)} e^{2\nu t} \, dt + \int_0^{\infty} \|u(t)\|^2_{L^2(M,E)} e^{2\gamma t} \, dt \right\}^{1/2}.$$

In particular, $L^2_{0\gamma}(M \times \mathbb{R}, E) = L^2_\gamma(M \times \mathbb{R}, E)$. Symbolically, we represent the cylinder $M \times \mathbb{R}$ equipped with the space $L^2_{\nu\gamma}(M \times \mathbb{R}, E)$ as shown in Fig. 6.6.

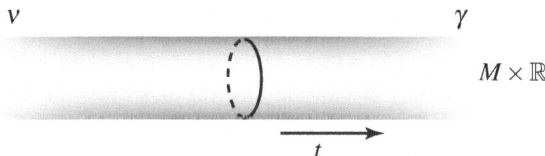

Figure 6.6: The cylinder $M \times \mathbb{R}$ with the space $L^2_{\nu\gamma}(M \times \mathbb{R}, E)$

Now the index increment in (ii) can be represented by the surgery shown in Fig. 6.7. This

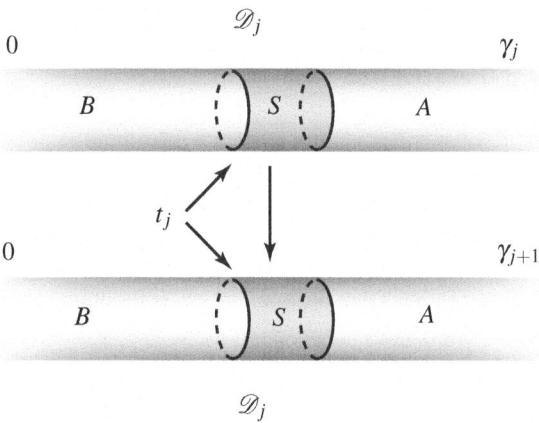

Figure 6.7: Relative index for γ changing from γ_j to γ_{j+1}

is clearly a surgery on the subset A. (The analytic expression for the operator itself is

6.3. Formula for the Spectral Flow

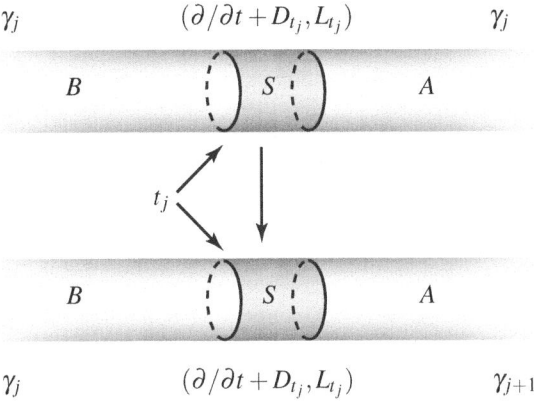

Figure 6.8: Relative index for γ changing from γ_j to γ_{j+1}: comparison operators

not changed; only the space on which the operator acts becomes different.) To compute this index increment, we modify our operators on the subset B. Namely, we replace the operator $\mathscr{D}_j$ by the operator $(\partial/\partial t + D_{t_j}, L_{t_j})$ with constant (independent of t) coefficients acting in the spaces as shown in Fig. 6.8.

Now everything is easy. The index of the upper operator is zero, because it is invertible. The index of the lower operator is $m_j \operatorname{sign}(\gamma_j - \gamma_{j+1})$. (Hint: the elements of the null space of this operator have the form $e^{-\lambda t} u_\lambda$, where u_λ is the eigenvector of D_t corresponding to the eigenvalue λ, and those of the adjoint operator have the form $e^{\lambda t} u_\lambda$; it remains to count how many of these fit into the space $L^2_{v\gamma}(M \times \mathbb{R}, E)$ and into the dual space $L^2_{-v, -\gamma}(M \times \mathbb{R}, E)$, respectively.) An application of Theorem 0.10 completes the proof of Theorem 6.6.

Remark 6.7. As we have already mentioned, we omit plenty of technical details. See [38] for these details.

6.4 Computation of the Spectral Flow for a Graphene Sheet

Now let us return to Example 6.4. For this example, we have the following corollary of Theorem 6.6.

Corollary 6.8. *For the spectral flow in Example 6.4, one has the formula*

$$\mathrm{sf}\{A_t, L_t\} = \sum_{j}^{l} wind_j(U),$$

where the sum is taken over those components of the boundary on which the function B in the boundary condition (6.2) *is positive.*

(This assertion was proved for the case of a single hole and conjectured for the case of arbitrarily many holes in [67].)

Indeed, if the sheet has only one hole, then it is homotopic to an annulus, and the desired formula follows from the results in Section 6.2. If the sheet has no holes at all, then the scalar bundle automorphism U is homotopic to identity, and the spectral flow is zero. Theorem 6.6 says that the spectral flow is the index of some operator on $X \times S_1$. It follows by the relative index superposition theorem applied to operators on $X \times S^1$ that we can cut our sheet into pieces each of which has at most one hole, and the spectral flow for the original sheet is just the sum of spectral flows for the resulting sheets. It would be an instructive exercise for the reader to write out the corresponding surgery diagrams.

Hint: by the relative index superposition theorem, if we cut a graphene sheet with (say) two holes into two sheets with one hole each, then the spectral flow increment is the same as if we cut a sheet without holes into two sheets without holes. But the spectral flow in the latter case is zero and does not change under surgery. Thus, we are done.

Bibliography

[1] Agranovich, M.: Elliptic boundary problems. In: M.S. Agranovich, Y.V. Egorov, M.A. Shubin (eds.) Partial Differential Equations IX. Elliptic Boundary Value Problems, no. 79 in Encyclopaedia of Mathematical Sciences, pp. 1–144. Springer Verlag, Berlin–Heidelberg (1997)

[2] Agranovich, M.S., Dynin, A.S.: General boundary-value problems for elliptic systems in higher-dimensional regions. Dokl. Akad. Nauk SSSR **146**, 511–514 (1962)

[3] Anghel, N.: An abstract index theorem on non-compact Riemannian manifolds. Houston J. of Math. **19**, 223–237 (1993)

[4] Atiyah, M.: Global theory of elliptic operators. In: Proc. of the Int. Symposium on Functional Analysis, pp. 21–30. University of Tokyo Press, Tokyo (1969)

[5] Atiyah, M., Bott, R.: The index problem for manifolds with boundary. In: Bombay Colloquium on Differential Analysis, pp. 175–186. Oxford University Press, Oxford (1964)

[6] Atiyah, M., Patodi, V., Singer, I.: Spectral asymmetry and Riemannian geometry I. Math. Proc. Cambridge Philos. Soc. **77**, 43–69 (1975)

[7] Atiyah, M., Patodi, V., Singer, I.: Spectral asymmetry and Riemannian geometry III. Math. Proc. Cambridge Philos. Soc. **79**, 71–99 (1976)

[8] Atiyah, M.F., Singer, I.M.: The index of elliptic operators on compact manifolds. Bull. Amer. Math. Soc. **69**, 422–433 (1963)

[9] Ballmann, W., Brüning, J., Carron, G.: Index theorems on manifolds with straight ends. Compos. Math. **148**(6), 1897–1968 (2012)

[10] Bär, C., Ballmann, W.: Boundary value problems for elliptic differential operators of first order. In: In Memory of C. C. Hsiung—Lectures given at the JDG Symposium, Lehigh University, June 2010, *Surveys in Differential Geometry*, vol. 17, pp. 1–78. International Press, Somerville (Mass.) (2012). ArXiv:1101.1196v2 [math.DG]

[11] Baum, P., Douglas, R.G.: *K*-homology and index theory. In: R. Kadison (ed.) Operator Algebras and Applications, no. 38 in Proc. Symp. Pure Math, pp. 117–173. American Mathematical Society (1982)

[12] Berry, M.V., Mondragon, R.J.: Neutrino billiards: time-reversal symmetry-breaking without magnetic fields. Proc. Roy. Soc. London Ser. A **412**(1842), 53–74 (1987)

[13] Blackadar, B.: K-Theory for Operator Algebras. No. 5 in Mathematical Sciences Research Institute Publications. Cambridge University Press (1998). Second edition

[14] Booss-Bavnbek, B., Lesch, M., Phillips, J.: Unbounded Fredholm operators and spectral flow. Canad. J. Math. **57**(2), 225–250 (2005)

[15] Booß-Bavnbek, B., Lesch, M., Zhu, C.: The Calderón projection: new definition and applications. J. Geom. Phys. **59**(7), 784–826 (2009)

[16] Booß-Bavnbek, B., Wojciechowski, K.: Elliptic Boundary Problems for Dirac Operators. Birkhäuser, Boston–Basel–Berlin (1993)

[17] Borisov, N.V., Müller, W., Schrader, R.: Relative index theorems and supersymmetric scattering theory. Commun. Math. Phys. **114**, 475–513 (1988)

[18] Bott, R., Seeley, R.: Some remarks on the paper of Callias. Commun. Math. Phys. **62**, 235–245 (1978)

[19] Brown, L., Douglas, R., Fillmore, P.: Extensions of C^*-algebras and K-homology. Ann. Math. II **105**, 265–324 (1977)

[20] Brüning, J., Lesch, M.: On boundary value problems for Dirac type operators. I. Regularity and self-adjointness. J. Funct. Anal. **185**(1), 1–62 (2001)

[21] Bunke, U.: Relative index theory. J. Funct. Anal. **105**, 63–76 (1992)

[22] Bunke, U.: A K-theoretic relative index theorem and Callias-type Dirac operators. Math. Ann. **303**(2), 241–279 (1995)

[23] Calderón, A.P.: Boundary value problems for elliptic equations. Outlines of the Joint Soviet–American Symposium on Partial Differential Equations, Novosibirsk, pp. 303–304 (1963)

[24] Callias, C.: Axial anomalies and index theorems on open spaces. Comm. Math. Phys. **62**(3), 213–234 (1978)

[25] Callias, C.: Index theorems on open spaces. Commun. Math. Phys. **62**, 213–234 (1978)

[26] Connes, A.: Noncommutative Geometry. Academic Press Inc., San Diego, CA (1994)

[27] Dixmier, J.: Les C^*-algebres et leurs representations. Gauthier-Villars, Paris (1969)

[28] Donnelly, H.: Essential spectrum and the heat kernel. J. Funct. Anal. **75**, 362–381 (1987)

[29] Epstein, C., Melrose, R.: Contact degree and the index of Fourier integral operators. Math. Res. Lett. **5**(3), 363–381 (1998)

[30] Farsi, C.: An orbifold relative index theorem. J. Geom. Phys. **57**(8), 1653–1668 (2007)

[31] Gilkey, P.B.: Invariance Theory, the Heat Equation, and the Atiyah-Singer Index Theorem, second edn. Studies in Advanced Mathematics. CRC Press, Boca Raton, FL (1995)

[32] Gromov, M., Lawson Jr., H.B.: Positive scalar curvature and the Dirac operator on complete Riemannian manifolds. Publ. Math. IHES **58**, 295–408 (1983)

[33] Hasan, M.Z., Kane, C.L.: *Colloquium*: Topological insulators. Rev. Mod. Phys. **82**, 3045–3067 (2010)

[34] Higson, N., Roe, J.: On the coarse Baum–Connes conjecture. In: Novikov conjectures, index theorems and rigidity, Vol. 2 (Oberwolfach, 1993), *London Math. Soc. Lecture Note Ser.*, vol. 227, pp. 227–254. Cambridge Univ. Press, Cambridge (1995)

[35] Higson, N., Roe, J.: Analytic K-Homology. Oxford University Press, Oxford (2000)

[36] Hörmander, L.: The Analysis of Linear Partial Differential Operators. III. Springer-Verlag, Berlin Heidelberg New York Tokyo (1985)

[37] Julg, P.: Indice relatif et K-théorie bivariant de Kasparov. C. R. Acad. Sci., Paris, Ser. I **307**(6), 243–248 (1988)

[38] Katsnelson, M., Nazaikinskii, V.: The Aharonov–Bohm effect for massless Dirac fermions and the spectral flow of Dirac-type operators with classical boundary conditions. Theoret. and Math. Phys. **172**(3), 1263–1277 (2012)

[39] Katsnelson, M.I.: Graphene: Carbon in Two Dimensions. Cambridge University Press, Cambridge (2012)

[40] Kottke, C.: An index theorem of Callias type for pseudodifferential operators. Journal of K-theory: K-theory and its Applications to Algebra, Geometry, and Topology (2011). Available on CJO 2011, doi:10.1017/is010011014jkt132

[41] Leichtnam, E., Nest, R., Tsygan, B.: Local formula for the index of a Fourier integral operator. J. Differential Geom. **59**(2), 269–300 (2001)

[42] Lesch, M.: Differential Operators of Fuchs Type, Conical Singularities, and Asymptotic Methods, *Teubner–Texte zur Mathematik*, vol. 136. B. G. Teubner Verlag, Stuttgart–Leipzig (1997)

[43] Loya, P., Park, J.: On the gluing problem for the spectral invariants of Dirac operators. Adv. Math. **202**(2), 401–450 (2006)

[44] Maslov, V.P.: Théorie des Perturbations et Méthodes Asymptotiques. Dunod, Paris (1972). French transl. from the Russian 1965 edition

[45] Melrose, R.: Transformation of boundary problems. Acta Math. **147**, 149–236 (1981)

[46] Mishchenko, A., Shatalov, V., Sternin, B.: Lagrangian Manifolds and the Maslov Operator. Springer–Verlag, Berlin–Heidelberg (1990)

[47] Mishchenko, A.S., Fomenko, A.T.: The index of elliptic operators over C^*-algebras. Izv. Akad. Nauk SSSR Ser. Mat. **43**(4), 831–859, 967 (1979)

[48] Nazaikinskii, V., Savin, A., Schulze, B.W., Sternin, B.: Elliptic Theory on Singular Manifolds. CRC-Press, Boca Raton (2005)

[49] Nazaikinskii, V., Savin, A., Sternin, B.: Homotopy classification of elliptic operators on stratified manifolds. Doklady: Mathematics **73**(3), 407–411 (2006)

[50] Nazaikinskii, V., Savin, A., Sternin, B.: On the homotopy classification of elliptic operators on stratified manifolds. Izv. Math. **71**(6), 1167–1192 (2007)

[51] Nazaikinskii, V., Savin, A., Sternin, B.: On the homotopy classification of elliptic operators on manifolds with corners. Doklady: Mathematics **75**(2), 186–189 (2007)

[52] Nazaikinskii, V., Savin, A., Sternin, B.: Elliptic theory on manifolds with corners. II: Homotopy classification and K-homology. Burghelea, Dan (ed.) et al., C^*-algebras and Elliptic Theory II. Selected papers of the international conference, Będlewo, Poland, January 2006. Basel: Birkhäuser. Trends in Mathematics, 207-226 (2008)

[53] Nazaikinskii, V., Schulze, B.W., Sternin, B.: The index of Fourier integral operators on manifolds with conical singularities. Izv. Ross. Akad. Nauk Ser. Mat. **65**(2), 127–154 (2001). English transl.: Izv. Math. **65**(2), 329–355 (2001)

[54] Nazaikinskii, V., Schulze, B.W., Sternin, B., Shatalov, V.: Spectral boundary value problems and elliptic equations on singular manifolds. Differentsial'nye Uravneniya **34**(5), 695–708 (1998). English transl.: Differential Equations **34**(5), 696–710 (1998)

[55] Nazaikinskii, V., Sternin, B.: Localization and surgery in index theory of elliptic operators. In: Conference: Operator Algebras and Asymptotics on Manifolds with Singularities, pp. 27–28. Stefan Banach International Mathematical Center, Universität Potsdam, Institut für Mathematik, Warsaw (1999)

[56] Nazaikinskii, V., Sternin, B.: Surgery and the Relative Index in Elliptic Theory. Univ. Potsdam, Institut für Mathematik, Potsdam (1999). Preprint No. 99/17

[57] Nazaikinskii, V., Sternin, B.: Localization and surgery in the index theory of elliptic operators. Russian Math. Dokl. **370**(1), 19–23 (2000)

[58] Nazaikinskii, V., Sternin, B.: A remark on elliptic theory on manifolds with isolated singularities. Dokl. Ross. Akad. Nauk **374**(5), 606–610 (2000)

[59] Nazaikinskii, V., Sternin, B.: On the local index principle in elliptic theory. Funkt. Anal. Prilozh. **35**(2), 37–52 (2001). English transl.: Funct. Anal. Appl. **35**(2), 111–123 (2001)

Bibliography

[60] Nazaikinskii, V., Sternin, B.: Surgery and the relative index in elliptic theory. In: Partial differential equations and spectral theory: PDE2000 Conference in Clausthal, Germany, *Operator Theory: Advances and Applications*, vol. 126, pp. 229–237. Birkhäuser, Basel, Boston, Berlin (2001)

[61] Nazaikinskii, V., Sternin, B.: Surgery and the relative index in elliptic theory. Abstr. Appl. Anal. (2006). Doi:10.1155/AAA/2006/98081

[62] Nazaikinskii, V.E.: On a KK-theoretic counterpart of relative index theorems (2012). (To appear in Russ. J. Math. Phys.) arXiv:1307.2770 [math.KT]

[63] Nazaikinskii, V.E.: Relative index theorem in K-homology (2012). (To appear in Funkts. Anal. Prilozh.) arXiv:1307.2833 [math.KT]

[64] Palais, R.S.: Seminar on the Atiyah–Singer Index Theorem. Princeton Univ. Press, Princeton, NJ (1965)

[65] Park, J., Wojciechowski, K.P.: Agranovich–Dynin formula for the zeta-determinants of the Neumann and Dirichlet problems. In: Spectral geometry of manifolds with boundary and decomposition of manifolds, *Contemp. Math.*, vol. 366, pp. 109–121. Amer. Math. Soc., Providence, RI (2005)

[66] Pedersen, G.K.: C^*-Algebras and Their Automorphism Groups, *London Mathematical Society Monographs*, vol. 14. Academic Press, London–New York (1979)

[67] Prokhorova, M.: The spectral flow for Dirac operators on compact planar domains with local boundary conditions. Comm. Math. Phys. pp. 1–30 (2013)

[68] Qi, X.L., Zhang, S.C.: Topological insulators and superconductors. Rev. Mod. Phys. **83**, 1057–1110 (2011)

[69] Roe, J.: Exotic cohomology and index theory. Bull. Amer. Math. Soc. (N.S.) **23**(2), 447–453 (1990)

[70] Roe, J.: A note on the relative index theorem. Quart. J. Math. Oxford Ser. (2) **42**(167), 365–373 (1991)

[71] Roe, J.: Notes on surgery and C^*-algebras. Sci. Bull. Josai Univ. (Special issue 2), 137–144 (1997). Surgery and geometric topology (Sakado, 1996)

[72] Roe, J.: Positive curvature, partial vanishing theorems, and coarse indices. arXiv:1210.6100 [math.KT] (2012)

[73] Roe, J., Siegel, P.: Sheaf theory and Paschke duality. arXiv:1210.6420 [math.KT] (2012)

[74] Rudin, W.: Functional Analysis, 2d edn. McGraw-Hill, New York (1991)

[75] Savin, A.: Elliptic operators on singular manifolds and K-homology. K-theory **34**(1), 71–98 (2005)

[76] Schulze, B.W., Sternin, B., Shatalov, V.: On general boundary value problems for elliptic equations. Math. Sb. **189**(10), 145–160 (1998). English transl.: Sbornik: Mathematics **189**, N 10 (1998), p. 1573–1586

[77] Schulze, B.W., Sternin, B., Shatalov, V.: On the index of differential operators on manifolds with conical singularities. Annals of Global Analysis and Geometry **16**(2), 141–172 (1998)

[78] Seeley, R.T.: Topics in pseudodifferential operators. In: L. Nirenberg (ed.) Pseudo-Differential Operators, pp. 167–305. C.I.M.E. Conference on pseudo-differential operators, Stresa 1968, Cremonese, Roma (1969)

[79] Shubin, M.A.: Pseudodifferential Operators and Spectral Theory. Nauka, Moscow (1978). English transl.: Springer-Verlag, Berlin–Heidelberg (1985)

[80] Teleman, N.: The index of signature operators on Lipschitz manifolds. Publ. Math. IHES **58**, 39–78 (1984)

[81] Weinstein, A.: Fourier integral operators, quantization, and the spectra of riemannian manifolds. In: Géométrie symplectique et physique mathématique (Aix-en-Provence, 1974), no. 237 in Colloque Internationale de CNRS, pp. 289–298 (1976)

[82] Weinstein, A.: Some questions about the index of quantized contact transformations. RIMS Kôkûryuku **104**, 1–14 (1977)

[83] Xie, Z.: Relative index pairing and odd index theorem for even dimensional manifolds. J. Funct. Anal. **260**(7), 2064–2085 (2011)

[84] Xie, Z., Yu, G.: A relative higher index theorem, diffeomorphisms and positive scalar curvature. arXiv:1204.3664 [math.KT] (2012)

[85] Yu, G.: Baum–Connes conjecture and coarse geometry. K-Theory **9**(3), 223–231 (1995)

[86] Yu, G.: Coarse Baum–Connes conjecture. K-Theory **9**(3), 199–221 (1995)

[87] Yu, G.: K-theoretic indices of Dirac type operators on complete manifolds and the Roe algebra. K-Theory **11**(1), 1–15 (1997)

[88] Yu, G.: Localization algebras and the coarse Baum–Connes conjecture. K-Theory **11**(4), 307–318 (1997)

Index

Symbols

C^*-algebra 46
   -s, sheaves of 79
   representation of 46
   unital, partition of unity in 61
K-homology group
   of a manifold 41
   of an algebra 42

A

Agranovich theorem 13, 89
Agranovich–Dynin theorem 14, 87
algebra $\mathfrak{B}(H)$ 46
almost commuting operators 6
almost inverse 28
Anghel theorem 12, 16
Atiyah–Bott formula 92
Atiyah–Patodi–Singer problem 83

B

Berry–Mondragon boundary conditions 95
Bojarski cutting conjecture 90
Bojarski theorem 90
boundary value problem
   classical 13
   Fredholm property 13
   general 83
   on the cylinder 85
   spectral 84
   with symmetric conormal symbol 91
Brillouin zone 95

Bunke theorem 77

C

c-Fredholm operator 28
   -s, coinciding on a subset 33
Callias type operator 16
Clifford bundle 71
Clifford multiplication 71
clutching 10
coarse geometry 79
coarse index 16
collar 23
collar space 8, 21–23
   modification of 29
   surgery of 29
commutative diagram of surgeries 7
composition of supports 26
cone-degenerate operator 10
   conormal symbol 10
   interior principal symbol of 10
covariant differentiation 71
cutting and pasting *see* surgery

D

Dirac operator 12, 16
   generalized 71
   invertible at infinity 76
   positive at infinity 73
Dirac type operator 93
   compatible boundary condition for 94

E

electron–hole pair 96

elliptic operator   1, 2, 87, 89
    on noncompact Riemannian manifold   11
    principal symbol of   1
    with parameter   84
elliptic symbol   2

F

finiteness theorem   2
Fourier integral operator   8
Fréchet topology   21
Fredholm module
    -s agreeing on an ideal   50
    -s, cutting and pasting of   50
    -s, unitary equivalence of   47
    block matrix form of   48
    degenerate   47
    graded   46
    nondegenerate   47
    ungraded   46
Fredholm property   2, 6

G

gauge transformation   96
general elliptic operator   6, 29
graphene   95
graphene sheet   96
Gromov–Lawson theorem   12, 16, 71, 73
    generalized   76

H

Hilbert module   16, 60
Hilbert space   46
homotopy invariant   2, 96

I

ideal of compact operators   46
index   2
index formula
    for cone-degenerate operators   10
    for quantized canonical transformations   16
index increment   *see* relative index
index problem   3

K

Kasparov KK-theory   61
Kasparov module   60
    -s, cutting and pasting of   63
    -s, agreeing on an ideal   63
    -s, standard homotopy of   61
    -s, unitary equivalence of   61
    degenerate   60

L

Lipschitz manifold   16
local index density   5
local index formula   5
locality   26
locally compact operator   46

M

magnetic flux quantum   96
manifolds coinciding at infinity   11
manifolds with conical singularities   9
    elliptic operator on   10
modification   *see* surgery

N

noncommutative geometry   44
normalization   44, 54

O

operator
    B-compact   60
    -s, isospectral   95
    adjointable   60
    compact   6
    local   42
    rank 1   60
    supported in a subset   26

Index

P

partition of unity   27
Paschke duality   79
pre-Hilbert module   59
proper operator   28
pseudodifferential operator   28
pseudodifferential projection   83

Q

quantized canonical transformation   15

R

relative index   3, 89
    on orbifolds   16
    Roe's construction   79
relative index theorem
    for boundary value problem   13
    for Kasparov modules   67
    local   16
Riemannian manifold:complete   71

S

Shapiro–Lopatinskii condition   13, 87, 89
Sobolev space   2
    weighted   10
spectral flow   95
spinor bundle   72
stable homotopy classification   41
stable homotopy invariant   41
superposition principle   3, 5
    for K-homology   9, 52
    for KK-theory   9
    for boundary value problems   12
    for cone-degenerate operators   10
    for Kasparov modules   67
    proof of   34
superposition theorem   7, 8
support   7, 23
    of an operator   25
    properties   26
surgery   3
    of c-Fredholm operators   33
    of general elliptic operators   6
surgery diagram   31
    commuting   31
    of c-Fredholm operators   34
    of Sobolev spaces   32
suspension   10

W

Weitzenböck formula   72
winding number   96

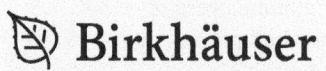

Pseudo-Differential Operators (PDO)
Theory and Applications

This is a series of moderately priced graduate-level textbooks and monographs appealing to students and experts alike. Pseudo-differential operators are understood in a very broad sense and include such topics as harmonic analysis, PDE, geometry, mathematical physics, microlocal analysis, time-frequency analysis, imaging and computations. Modern trends and novel applications in mathematics, natural sciences, medicine, scientific computing, and engineering are highlighted.

Edited by
M. W. Wong, York University, Canada
In cooperation with an international editorial board

■ **PDO 9: Cohen, L.**, The Weyl Operator and its Generalization (2013).
ISBN 978-3-0348-0293-2

The concept of associating ordinary functions with operators has arisen in many areas of science and mathematics, and up to the beginning of the twentieth century many isolated results were obtained. These developments were mostly based on associating a function of one variable with one operator, the operator generally being the differentiation operator. With the discovery of quantum mechanics in the years 1925–1930, there arose, in a natural way, the issue that one has to associate a function of two variables with a function of two operators that do not commute. Methods to do so became known as rules of association, correspondence rules, or ordering rules. This has led to a wonderfully rich mathematical development that has found applications in many fields. Subsequently it was realized that for every correspondence rule there is a corresponding phase-space distribution. Now the fields of correspondence rules and phase-space distributions are intimately connected. A similar development occurred in the field of time-frequency analysis where the aim is to understand signals with changing frequencies.
The Weyl Operator and Its Generalization aims at bringing together the basic results of the field in a unified manner. A wide audience is addressed, particularly students and researchers who want to obtain an up-to-date working knowledge of the field. The mathematics is accessible to the uninitiated reader and is presented in a straightforward manner.

■ **PDO 8: Unterberger, A.**, Pseudodifferential Analysis, Automorphic Distributions in the Plane and Modular Forms (2011).
ISBN 978-3-0348-0165-2

Pseudodifferential analysis, introduced in this book in a way adapted to the needs of number theorists, relates automorphic function theory in the hyperbolic half-plane Π to automorphic distribution theory in the plane. Spectral-theoretic questions are discussed in one or the other environment: in the latter one, the problem of decomposing automorphic functions in Π according to the spectral decomposition of the modular Laplacian gives way to the simpler one of decomposing automorphic distributions in $\mathbf{R}^2$ into homogeneous components. The Poincaré summation process, which consists in building automorphic distributions as series of g-transforms, for $g \in SL(2;\mathbf{Z})$, of some initial function, say in $S(\mathbf{R}^2)$, is analyzed in detail. On Π, a large class of new automorphic functions or measures is built in the same way: one of its features lies in an interpretation, as a spectral density, of the restriction of the zeta function to any line within the critical strip.

■ **PDO 7: de Gosson, M.**, Symplectic Methods in Harmonic Analysis and in Mathematical Physics (2011).
ISBN 978-3-7643-9991-7

The aim of this book is to give a rigorous and complete treatment of various topics from harmonic analysis with a strong emphasis on symplectic invariance properties, which are often ignored or underestimated in the time-frequency literature. The topics that are addressed include (but are not limited to) the theory of the Wigner transform, the uncertainty principle (from the point of view of symplectic topology), Weyl calculus and its symplectic covariance, Shubin's global theory of pseudo-differential operators, and Feichtinger's theory of modulation spaces. Several applications to time-frequency analysis and quantum mechanics are given, many of them concurrent with ongoing research.
This book is primarily directed towards students or researchers in harmonic analysis (in the broad sense) and towards mathematical physicists working in quantum mechanics. It can also be read with profit by researchers in time-frequency analysis, providing a valuable complement to the existing literature on the topic. A certain familiarity with Fourier analysis and introductory functional analysis (e.g. the elementary theory of distributions) is assumed. Otherwise, the book is largely self-contained and includes an extensive list of references.

The manufacturer's authorised representative in the EU is Springer Nature Customer Service Centre GmbH, Europaplatz 3, 69115 Heidelberg, Germany. If you have any concerns regarding our products, please contact ProductSafety@springernature.com

Printed and bound by CPI Group (UK) Ltd, Croydon, CR0 4YY

23/03/2026

02076393-0018